THE FUTURE OF LIFE

THE FUTURE OF LIFE

Edward O. Wilson

ALFRED A. KNOPF

NEW YORK

2002

THIS IS A BORZOI BOOK
PUBLISHED BY ALFRED A. KNOPF

Copyright © 2002 by E. O. Wilson

Published in the United States by Alfred A. Knopf, a division of Random House, Inc.,
New York, and simultaneously in Canada by Random House of Canada Limited,
Toronto. Distributed by Random House, Inc., New York.
www.aaknopf.com

Knopf, Borzoi Books, and the colophon are registered trademarks of
Random House, Inc.

Chapter 2 of this work was originally published in *Scientific American,*
and Chapter 5 in the *Wilson Quarterly.*

Library of Congress Cataloging-in-Publication Data
Wilson, Edward Osborne, 1929–
The future of life / Edward O. Wilson. — 1st ed.
p.   cm.
ISBN 0-679-45078-5 (hardcover)
1. Endangered species. 2. Nature conservation.
3. Environmental degradation. I. Title.
QH75 .W535 2002
333.95'22—dc21   2001038316

Diagram representing endangered and extinct species and races by Isabella Kirkland.

Manufactured in the United States of America

First Edition

———————O———————

*In the end, our society will be defined
not only by what we create, but by what
we refuse to destroy.*

—John C. Sawhill (1936–2000), president,
The Nature Conservancy, 1990–2000

CONTENTS

ENDANGERED AND EXTINCT SPECIES AND RACES

(represented in jacket art, and as numbered in adjacent diagram)

1. Hawksbill sea turtle—*Eretmochelys imbricata*
2. Condor egg—*Gymnogyps californianus*
3. Giant kangaroo rat—*Dipodomys ingens*
4. Little Kern golden trout—*Oncorhynchus aquabonita whitei*
5. San Francisco garter snake—*Thamnophis sirtalis tetrataenia*
6. Golden toad—*Bufo periglenes*
7. I'iwi—*Vestiaria coccinea*
8. Okeechobee gourd—*Cucurbita okeechobeensis*
9. Presidio manzanita—*Arctostaphylos pungens* var. *ravenii*
10. James spineymussel—*Pleurobema collina*
11. Fat pocketbook pearly mussel—*Potamilus capax*
12. Dwarf wedge mussel—*Alasmidonta heterodon*
13. Geyser's panicum—*Dichanthelium lanuginosum* var. *thermale*
14. Valley oak—*Quercus lobata*
15. Guadalupe violet—*Viola guadalupensis*
16. Missouri bladderpod—*Lesquerella filiformis*
17. Indian Knob mountainbalm—*Eriodictyon altissimum*
18. American burying beetle—*Nicrophorus americanus*
19. Vine Hill clarkia—*Clarkia imbricata*
20. Price's potato bean—*Apios priceana*
21. Na'u—*Gardenia brighamii*
22. Valley elderberry longhorn beetle—*Desmocerus californicus dimorphus*
23. Baker's blennosperm—*Blennosperma bakeri*

---O---

PROLOGUE

A LETTER TO THOREAU

Henry!

May I call you by your Christian name? Your words invite familiarity and make little sense otherwise. How else to interpret your insistent use of the first personal pronoun? *I* wrote this account, you say, here are *my* deepest thoughts, and no third person placed between us could ever be so well represented. Although *Walden* is sometimes oracular in tone, I don't read it, the way some do, as an oration to the multitude. Rather, it is a work of art, the testament of a citizen of Concord, in New England, from one place, one time, and one writer's personal circumstance that manages nevertheless to reach across five generations to address accurately the general human condition. Can there be a better definition of art?

You brought me here. Our meeting could have just as well been a woodlot in Delaware, but here I am at the site of your cabin on the edge of Walden Pond. I came because of your stature in literature and the conservation movement, but also—less nobly, I confess—because my home is in Lexington, two towns over. My pilgrimage is a pleasant afternoon's excursion to a nature reserve. But mostly I came because of all your contemporaries you are the one I most need to understand. As a biologist with a modern scientific library, I know more than Darwin knew. I can imagine the measured responses of that country gentleman to a voice a century and a half beyond his own. It is not a satisfying fantasy: the Victorians have for the most part settled into a comfortable corner of our remembrance. But I cannot imagine your responses, at least not all of them. Too many shadowed residues there in your text, too many emotional trip wires. You left too soon, and your restless spirit haunts us still.

Is it so odd to speak apostrophically across 150 years? I think not. Certainly not if the subject is natural history. The wheels of organic evolution turn at a millennial pace, too slowly for evolution to have transformed species from your time to mine. The natural habitats they compose also remain mostly unchanged. Walden Woods around the pond, having been only partly cut and never plowed, looks much the same in my time as in yours, although now more fully wooded. Its ambience can be expressed in similar language.

Anyway, the older I become, the more it makes sense to measure history in units of life span. That pulls us closer together in real time. Had you lived to eighty instead of just forty-four, we might today have a film clip of you walking on Walden Pond beach through a straw-hatted and parasoled crowd on holiday. We could listen to your recorded voice from one of Mr. Edison's wax cylinders. Did you speak with a slight burr, as generally believed? I am seventy-two now, old enough to have had tea with Darwin's last surviving granddaughter at the University of Cambridge. While a Harvard graduate student I discussed my first articles on evolution with Julian Huxley, who as a little boy sat on the knee of his grandfather Thomas Henry Huxley, Darwin's "bulldog" disciple and personal friend. You will know what I am talking about. You still had three years to live when in 1859 *The Origin of Species* was published. It was the talk of Harvard and salons along the Atlantic seaboard. You purchased one of the first copies available in America and annotated it briskly. And here is one more circumstance on which I often reflect: as a child I could in theory have spoken to old men who visited you at Walden Pond when they were children of the same age. Thus only one living memory separates us. At the cabin site even that seems to vanish.

Forgive me, I digress. I am here for a purpose: to become more a Thoreauvian, and with that perspective better to explain to you, and in reality to others and not least to myself, what has happened to the world we both have loved.

The landscape away from Walden Pond, to start, has changed drastically. In your time the forest was almost gone. The tallest white pines had been cut long before and hauled away to Boston to be trimmed into ship masts. Other timber was harvested for houses, railroad ties, and fuel. Most of the swamp cedars had become roof shingles. America, still a wood-powered nation, was approaching its first energy crisis as charcoal and cordwood ran short. Soon everything would change. Then coal would fill the breach and catapult the industrial revolution forward at an even more furious pace.

When you built your little house from the dismantled planks of James Collins's shanty in 1845, Walden Woods was a threatened oasis in a mostly treeless terrain. Today it is pretty much the same, although forest has grown up to fill the farmland around it. The trees are still scraggly second-growth descendants of the primeval giants that clothed the lake banks until the mid-1700s. Around the cabin site, beech, hickory, red maple, and scarlet and white oak push up among half-grown white pines in a bid to reestablish the rightful hardwood domination of southern New England forests. Along the path from your cabin on down to the nearest inlet—now called Thoreau's Cove—these trees give way to an open stand of larger white pines, whose trunks are straight and whose branches are evenly spread and high off the ground. The undergrowth consists of a sparse scattering of saplings and huckleberry. The American chestnut, I regret to report, is gone, done in by an overzealous European fungus. Only a few sprouts still struggle up from old stumps here and there, soon to be discovered by the fungus and killed back. Sprouting their serrate leaves, the doomed saplings are faint reminders of the mighty species that once composed a full quarter of the eastern virgin forest. But all the other trees and shrubs you knew so well still flourish. The red maple is more abundant than in your day. It is more than ever both the jack-of-all-trades in forest regeneration and the crimson glory of the New England autumn.

I can picture you clearly as your sister Sophia sketched you, sitting here on the slightly raised doorstep. It is a cool morning in June, by my tastes the best month of the year in New England. In my imagination I have settled beside you. We gaze idly across this spring-fed lake of considerable size that New Englanders perversely call a pond. Today in this place we speak a common idiom, breathe the same clean air, listen to the whisper of the pines. We scuff the familiar leaf litter with our shoes, pause, look up to watch a circling red-tailed hawk pass overhead. Our talk drifts from here to there but never so far from natural history as to break the ghostly spell and never so intimate as to betray the childish sources of our common pleasure. A thousand years will pass and Walden Woods will stay the same, I think, a flickering equilibrium that works its magic on human emotion in variations with each experience.

We stand up to go a-sauntering. We descend the cordwood path to the lake shore, little changed in contour from the sketch you made in 1846, follow it around, and coming to a rise climb to the Lincoln Road, then circle back to the Wyman Meadow and on down to Thoreau's Cove, completing a round-trip of two miles. We search along the way for the woods least savaged by axe and crosscut saw. It is our intention to work not around but *through* these remnants. We stay within a quarter-mile or so of the lake, remembering that in your time almost all the land outside the perimeter woods was cultivated.

Mostly we talk in alternating monologues, because the organisms we respectively favor are different enough to require cross-explanation. There are two kinds of naturalists, you will agree, defined by the search images that guide them. The first—your tribe—are intent on finding big organisms: plants, birds, mammals, reptiles, amphibians, perhaps butterflies. Big-organism people listen for animal calls, peer into the canopy, poke into tree hollows, search mud banks for scat and spoor. Their line of sight vacillates around the horizontal, first upward to scan the canopy, then down to peer at the ground. Big-organism people search for a single find good enough for the day. You, I recall, thought little of walking four miles or more to see if a certain plant had begun to flower.

I am a member of the other tribe—a lover of little things, a hunter also, but more the snuffling opossum than the questing panther. I think in millimeters and minutes, and am nowhere near patient as I prowl, having been spoiled forever by the richness of invertebrates and quick reward for little effort. Let me enter a tract of rich forest and I seldom walk more than a few hundred feet. I halt before the first promising rotten log I encounter. Kneeling, I roll it over, and always there is instant gratification from the little world hidden beneath. Rootlets and fungal strands pull apart, adhering flakes of bark fall back to earth. The sweet damp musty scent of healthy soil rises like a perfume to the nostrils that love it. The inhabitants exposed are like deer jacklighted on a country road, frozen in a moment of their secret lives. They quickly scatter to evade the light and desiccating air, each maneuvering in the manner particular to its species. A female wolf spider sprints headlong for several body lengths and, finding no shelter, stops and stands rigid. Her brindled integument provides camouflage, but the white silken egg case she carries between her pedipalps and fangs gives her away. Close by, julid millipedes, which were browsing on mold when the cataclysm struck, coil their bodies in defensive spirals. At the far end of the exposed surface a large scolopendrid centipede lies partly concealed beneath decayed bark fragments. Its sclerites are a glistening brown armor, its jaws poison-filled hypodermic needles, its legs downward-curving scythes. The scolopendrid offers no threat unless you pick it up.

But who would dare touch this miniature dragon? Instead I poke it with the tip of a twig. *Get out of there!* It writhes, spins around, and is gone in a flash. Now I can safely rake my fingers through the humus in search of less threatening species.

These arthropods are the giants of the microcosm (if you will allow me to continue what has turned into a short lecture). Creatures their size are present in dozens—hundreds, if an ant or termite colony is present. But these are comparatively trivial numbers. If you focus down by a power of ten in size, enough to pick out animals barely visible to the naked eye, the numbers jump to thousands. Nematode and enchytraeid pot worms, mites, springtails, pauropods, diplurans, symphylans, and tardigrades seethe in the underground. Scattered out on a white ground cloth, each crawling speck becomes a full-blown animal. Together they are far more striking and diverse in appearance than snakes, mice, sparrows, and all the other vertebrates hereabouts combined. Their home is a labyrinth of miniature caves and walls of rotting vegetable debris cross-strung with ten yards of fungal threads. And they are just the surface of the fauna and flora at our feet. Keep going, keep magnifying until the eye penetrates microscopic water films on grains of sand, and there you will find ten billion bacteria in a thimbleful of soil and frass. You will have reached the energy base of the decomposer world as we understand it 150 years after your sojourn in Walden Woods.

Untrammeled nature exists in the dirt and rotting vegetation beneath our shoes. The wilderness of ordinary vision may have vanished—wolf, puma, and wolverine no longer exist in the tamed forests of Massachusetts. But another, even more ancient wilderness lives on. The microscope can take you there. We need only narrow the scale of vision to see a part of these woods as they were a thousand years ago. This is what, as a small-organism naturalist, I can tell you.

"Thó-reau." Your family put the emphasis on the first syllable, as in "thorough," did it not? At least that is what your close friend Ralph Waldo Emerson scribbled on a note found among his papers. Thoreau, thorough naturalist, you would have liked the Biodiversity Day we held in your honor here recently. It was conceived by Peter Alden, a Concord resident and international wildlife tour guide. (Easy name to remember; he is a descendant of John Alden of Pilgrim fame.) On July 4, 1998, the anniversary of the day in 1845 you moved furniture into the Walden cabin, Peter and I were joined by more than a hundred other naturalists from around New England. We set out to list all the wild species of organisms—plants, animals, and fungi—we could find in one day with unaided vision or hand lens within a broad section of Concord and Lincoln around Walden Pond. We aimed for a thousand. The final tally, announced to the thorn-scratched, mosquito-bitten group assembled at an outdoor meal that evening, was 1,904. Well, actually 1,905, to stretch the standards a bit, because the next day a moose *(Alces alces)* came from somewhere and strolled into Concord Center. It soon strolled out again, and evidently departed the Concord area, thus lowering the biodiversity back to the July 4 level.

If you could have come back that Biodiversity Day you might have joined us unnoticed (that is, if you refrained from bringing up President Polk and the Mexican question). Even the 1840s clothing would not have betrayed you, given our own scruffy and eclectic field wear. You would have understood our purpose too. From your last two books, *Faith in a Seed* and *Wild Fruits* (finally rescued from your almost indecipherable notes and published in the 1990s), it is apparent that you were moving toward scientific natural history when your life prematurely ended. It was logical for you to take that turn: the beginning of every science is the description and naming of phenomena. Human beings seem to have an instinct to master their surroundings that way. We cannot think clearly about a plant or animal until we have a name for it; hence the pleasure of bird watching with a field guide in hand. Alden's idea quickly caught on. As I write, in 2001, Biodiversity Days, or "bioblitzes" as they are also called, are being held or planned elsewhere in the United States as well as in Austria, Germany, Luxembourg, and Switzerland. In June 2001 we were joined for a third event in Massachusetts by students from 260 towns over the entire state.

At Walden Pond that first day I met Brad Parker, one of the character actors who play you while giving tours around the reconstructed cabin. He is steeped in Thoreauviana, and eerily convincing. He refused to deviate even one second from your persona as we talked, bless him, and for a pleasant hour I lived in the virtual 1840s he created. Of course, to reciprocate I invited him to peer with me at insects and other invertebrates beneath nearby stones and fallen dead branches. We moved on to a clump of bright yellow mushrooms. Then Neo-Thoreau mentioned a singing wood thrush in the canopy above us, which my deafness in the upper registers prevented me from hearing. We went on like this for a while, with his making nineteenth-century sallies and responses and my struggling to play the part of a time-warped visitor. No mention was made of the thunder of aircraft above us on their approach to Hanscom Field. Nor did I think it anomalous that at sixty-nine I was speaking to a reanimation of you, Henry Real-Thoreau, at thirty. In one sense it was quite appropriate. The naturalists of my generation are you grown older and more knowledgeable, if not wiser.

A case in point on the growth of knowledge. Neo-Thoreau and I talked about the ant war you described in *Walden*. One summer day you found red ants locked in mandible-to-mandible combat with black ants all around your cabin. The ground was littered with the dead and dying, and the ambulatory maimed fought bravely on. It was an ant-world Austerlitz, as you said, a conflict dwarfing the skirmish on the Concord Bridge that started the American Revolution a rifle shot from Walden Pond. May I presume to tell you what you saw? It was a slave raid. The slavers were the red ants, most likely *Formica subintegra,* and the victims were the black ants, probably *Formica subsericea.* The red ants capture the infants of their victims, or more precisely, their cocoon-clad pupae. Back in the red-ant nest the kidnapped pupae complete their development and emerge from their cocoons as adult workers. Then, because they instinctively accept the first workers they meet as nestmates, they enter into voluntary servitude to their captors. Imagine that! A slave raid at the doorstep of one of America's most ardent abolitionists. For millions of years this harsh Darwinian strategy has prevailed, and so will it ever be, with no hope that a Lincoln, a Thoreau, or an Underground Railroad might arise in the formicid world to save the victim colonies.

Now, prophet of the conservation movement, mentor of Gandhi and Martin Luther King Jr., accept this tribute tardily given. Keen observer of the human condition, scourge of the philistine culture, Greek stoic adrift in the New World, you are reborn in each generation and vested with new meaning and nuance. Sage of Concord—Saint Henry, they sometimes call you—you've fairly earned your place in history.

On the other hand, you were not a great naturalist. (Forgive me!) Even had you kept entirely to natural history during your short life, you would have ranked well below William Bartram, Louis Agassiz, and that prodigious collector of North American plants John Torrey, and be scarcely remembered today. With longer life it would likely have been different, because you were building momentum in natural history rapidly when you left us. And to give you full credit, your ideas on succession and other properties of living communities pointed straight toward the modern science of ecology.

That doesn't matter now. I understand why you came to Walden Pond; your words are clear enough on that score. Granted, you chose this spot primarily to study nature. But you could have done that as easily and far more comfortably on daily excursions from your mother's house in Concord Center, half an hour's walk away, where in fact you did frequently repair for a decent meal. Nor was your little cabin meant to be a wilderness hermitage. No wilderness lay within easy reach anyway, and even the woods around Walden Pond had shrunk to their final thin margins by the

1840s. You called solitude your favorite companion. You were not afraid, you said, to be left to the mercy of your own thoughts. Yet you craved humanity passionately, and your voice is anthropocentric in mood and philosophy. Visitors to the Walden cabin were welcomed. Once a group of twenty-five or more crowded into the solitary room of the tiny house, shoulder to shoulder. You were not appalled by so much human flesh pressed together (but I am). You were lonely at times. The whistle of a passing train on the Fitchburg track and the distant rumble of oxcarts crossing a bridge must have given you comfort on cold, rainy days. Sometimes you went out looking for someone, anyone, in spite of your notorious shyness, just to have a conversation. You fastened on them, as you put it, like a bloodsucker.

In short, you were far from the hard-eyed frontiersman bearing pemmican and a long rifle. Frontiersmen did not saunter, botanize, and read Greek. So how did it happen that an amateur naturalist perched in a toy house on the edge of a ravaged woodland became the founding saint of the conservation movement? Here is what I believe happened. Your spirit craved an epiphany. You sought enlightenment and fulfillment the Old Testament way, by reduction of material existence to the fundamentals. The cabin was your cave on the mountainside. You used poverty to purchase a margin of free existence. It was the only method you could devise to seek the meaning in a life otherwise smothered by quotidian necessity and haste. You lived at Walden, as you said (I dare not paraphrase),

to front only the essential facts of life, and see if I could not learn what it had to teach, and not, when I came to die, discover that I had not lived . . . to live deep and suck out all the marrow of life, to live so sturdily and Spartan-like as to put to rout all that was not life, to cut a broad swath and shave close, to drive life into a corner, and reduce it to its lowest terms, and, if it proved to be mean, why then to get the whole and genuine meanness of it, and publish its meanness to the world; or if it were sublime, to know it by experience, and be able to give a true account of it in my next excursion.

You were mistaken, I think, to suppose that there are as many ways of life possible as radii that can be drawn from the center of a circle, and your choice just one of them. On the contrary, the human mind can develop along only a very few pathways imaginable. They are selected by satisfactions we instinctively seek in common. The sturdiness of human nature is the reason people plant flowers, gods live on high mountains, and a lake is the eye of the world through which—your metaphor—we can measure our own souls.

It is exquisitely human to search for wholeness and richness of experience. When these qualities are lost among the distracting schedules of everyday life, we seek them elsewhere. When you stripped your outside obligations to the survivable minimum, you placed your trained and very active mind in an unendurable vacuum. And this is the essence of the matter: in order to fill the vacuum, you discovered the human proclivity to embrace the natural world.

Your childhood experience told you exactly where to go. It could not be a local cornfield or gravel pit. Nor the streets of Boston, which, however vibrant as the hub of a growing nation, might cost a layabout his dignity and even his life. It had to be a world both tolerant of poverty and rich and beautiful enough to be spiritually rewarding. Where around Concord could that possibly be but a woodlot next to a lake?

You traded most of the richness of social existence for an equivalent richness of the natural world. The choice was entirely logical, for the following reason. Each of us finds a comfortable position somewhere along the continuum that ranges from complete withdrawal and self-absorption at one end to full civic engagement and reciprocity at the other. The position is never fixed. We fret, vacillate, and steer our lives through the riptide of countervailing instincts that press from both ends of the continuum. The uncertainty we feel is not a curse. It is not a confusion on the road out of Eden. It is just the human condition. We are intelligent mammals, fitted by evolution—by God, if you prefer—to pursue personal ends through cooperation. Our priceless selves and family first, society next. In this respect we are the polar opposite of your cabinside ants, bound together as replaceable parts of a superorganism. Our lives are therefore an insoluble problem, a dynamic process in search of an indefinable goal. They are neither a celebration nor a spectacle but rather, as a later philosopher put it, a predicament. Humanity is the species forced by its basic nature to make moral choices and seek fulfillment in a changing world by any means it can devise.

You searched for essence at Walden and, whether successful in your own mind or not, you hit upon an ethic with a solid feel to it: nature is ours to explore forever; it is our crucible and refuge; it is our natural home; it is all these things. Save it, you said: in wildness is the preservation of the world.

Now, in closing this letter, I am forced to report bad news. (I put it off till the end.) The natural world in the year 2001 is everywhere disappearing before our eyes—cut to pieces, mowed down, plowed under, gobbled up, replaced by human artifacts.

No one in your time could imagine a disaster of this magnitude. Little more than a billion people were alive in the 1840s. They were overwhelmingly agricultural, and few families needed more than two or three acres to survive. The American frontier was still wide open. And far away on continents to the south, up great rivers, beyond unclimbed mountain ranges, stretched unspoiled equatorial forests brimming with the maximum diversity of life. These wildernesses seemed as unattainable and timeless as the planets and stars. That could not last, because the mood of Western civilization is Abrahamic. The explorers and colonists were guided by a biblical prayer: May we take possession of this land that God has provided and let it drip milk and honey into our mouths, forever.

Now, more than six billion people fill the world. The great

majority are very poor; nearly one billion exist on the edge of starvation. All are struggling to raise the quality of their lives any way they can. That unfortunately includes the conversion of the surviving remnants of the natural environment. Half of the great tropical forests have been cleared. The last frontiers of the world are effectively gone. Species of plants and animals are disappearing a hundred or more times faster than before the coming of humanity, and as many as half may be gone by the end of this century. An Armageddon is approaching at the beginning of the third millennium. But it is not the cosmic war and fiery collapse of mankind foretold in sacred scripture. It is the wreckage of the planet by an exuberantly plentiful and ingenious humanity.

The race is now on between the technoscientific forces that are destroying the living environment and those that can be harnessed to save it. We are inside a bottleneck of overpopulation and wasteful consumption. If the race is won, humanity can emerge in far better condition than when it entered, and with most of the diversity of life still intact.

The situation is desperate—but there are encouraging signs that the race can be won. Population growth has slowed, and, if the present trajectory holds, is likely to peak between eight and ten billion people by century's end. That many people, experts tell us, can be accommodated with a decent standard of living, but just barely: the amount of arable land and water available per person, globally, is already declining. In solving the problem, other experts tell us, it should also be possible to shelter most of the vulnerable plant and animal species.

In order to pass through the bottleneck, a global land ethic is urgently needed. Not just any land ethic that might happen to enjoy agreeable sentiment, but one based on the best understanding of ourselves and the world around us that science and technology can provide. Surely the rest of life matters. Surely our stewardship is its only hope. We will be wise to listen carefully to the heart, then act with rational intention and all the tools we can gather and bring to bear.

Henry, my friend, thank you for putting the first element of that ethic in place. Now it is up to us to summon a more encompassing wisdom. The living world is dying; the natural economy is crumbling beneath our busy feet. We have been too self-absorbed to foresee the long-term consequences of our actions, and we will suffer a terrible loss unless we shake off our delusions and move quickly to a solution. Science and technology led us into this bottleneck. Now science and technology must help us find our way through and out.

You once said that old deeds are for old people, and new deeds are for new. I think that in historical perspective it is the other way around. You were the new and we are the old. Can we now be the wiser? For you, here at Walden Pond, the lamentation of the mourning dove and the green frog's *t-r-r-oonk!* across the predawn water were the true reason for saving this place. For us, it is an exact knowledge of what that truth is, all that it implies, and how to employ it to best effect. So, two truths. We will have them both, you and I and all those now and forever to come who accept the stewardship of nature.

Affectionately yours,
Edward

THE FUTURE OF LIFE

CHAPTER 1

———o———

TO THE ENDS OF EARTH

The totality of life, known as the biosphere to scientists and creation to theologians, is a membrane of organisms wrapped around Earth so thin it cannot be seen edgewise from a space shuttle, yet so internally complex that most species composing it remain undiscovered. The membrane is seamless. From Everest's peak to the floor of the Mariana Trench, creatures of one kind or another inhabit virtually every square inch of the planetary surface. They obey the fundamental principle of biological geography, that wherever there is liquid water, organic molecules, and an energy source, there is life. Given the near-universality of organic materials and energy of some kind or other, water is the deciding element on planet Earth. It may be no more than a transient film on grains of sand, it may never see sunlight, it may be boiling hot or supercooled, but there will be some kind of organism living in or upon it. Even if nothing alive is visible to the naked eye, single cells of microorganisms will be growing and reproducing there, or at least dormant and awaiting the arrival of liquid water to kick them back into activity.

An extreme example is the McMurdo Dry Valleys of Antarctica, whose soils are the coldest, driest, and most nutritionally deficient in the world. On first inspection the habitat seems as sterile as a cabinet of autoclaved glassware. In 1903, Robert F. Scott, the first to explore the region, wrote, "We have seen no living thing, not even a moss or lichen; all that we did find, far inland among the moraine heaps, was the skeleton of a Weddell seal, and how that came there is beyond guessing." On all of Earth the McMurdo Dry Valleys most resemble the rubbled plains of Mars.

But the trained eye, aided by a microscope, sees otherwise. In the parched streambeds live twenty species of photosynthetic bacteria, a comparable variety of mostly single-celled algae, and an array of microscopic invertebrate animals that feed on these primary producers. All depend on the summer flow of glacial and icefield meltwater for their annual spurts of growth. Because the paths of the streams change over time, some of the populations are stranded and forced to wait for years, perhaps centuries, for the renewed flush of meltwater. In the even more brutal conditions on bare land away from the stream channels live sparse assemblages of microbes and fungi together with rotifers, bear animalcules, mites, and springtails feeding on them. At the top of this rarefied food web are four species of nematode worms, each specialized to consume different species in the rest of the flora and fauna. With the mites and springtails they are also the largest of the animals, McMurdo's equivalent of elephants and tigers, yet all but invisible to the naked eye.

The McMurdo Dry Valleys's organisms are what scientists call extremophiles, species adapted to live at the edge of biological tolerance. Many populate the environmental ends of Earth, in places that seem uninhabitable to gigantic, fragile animals like ourselves. They constitute, to take a second example, the "gardens" of the Antarctic sea ice. The thick floes, which blanket millions of square miles of ocean water around the continent much of the year, seem forbiddingly hostile to life. But they are riddled with channels of slushy brine in which single-celled algae flourish year-round, assimilating the carbon dioxide, phosphates, and other nutrients that work up from the ocean below. The garden photosynthesis is driven by energy from sunlight penetrating the translucent matrix. As the ice melts and erodes during the polar summer, the algae sink into the water below, where they are consumed by cope-pods and krill. These tiny crustaceans in turn are the prey of fish whose blood is kept liquid by biochemical antifreezes.

The ultimate extremophiles are certain specialized microbes, including bacteria and their superficially similar but genetically very different relatives the archaeans. (To take a necessary digres-sion: biologists now recognize three domains of life on the basis of DNA sequences and cell structure. They are the Bacteria, which are the conventionally recognized microbes; the Archaea, the other microbes; and the Eukarya, which include the single-celled protists or "protozoans," the fungi, and all of the animals, includ-ing us. Bacteria and archaeans are more primitive than other organisms in cell structure: they lack membranes around their nuclei as well as organelles such as chloroplasts and mitochon-dria.) Some specialized species of bacteria and archaeans live in the walls of volcanic hydrothermal vents on the ocean floor, where they multiply in water close to or above the boiling point. A bac-terium found there, *Pyrolobus fumarii,* is the reigning world cham-pion among the hyperthermophiles, or lovers of extreme heat. It can reproduce at 235°F, does best at 221°F, and stops growing when the temperature drops to a chilly 194°F. This extraordinary feat has prompted microbiologists to inquire whether even more advanced, ultrathermophiles exist, occupying geothermal waters at 400°F or even higher. Watery environments with temperatures that hot exist. The submarine spumes close to the *Pyrolobus fumarii* bacterial colonies reach 660°F. The absolute upper limit of life as a whole, bacteria and archaeans included, is thought to be about 300°F, at which point organisms cannot sustain the integrity of DNA and the proteins on which known forms of life depend. But until the search for ultrathermophiles, as opposed to mere hyperthermophiles, is exhausted, no one can say for certain that these intrinsic limits actually exist.

During more than three billion years of evolution, the bacteria and archaeans have pushed the boundaries in other dimensions of physiological adaptation. One species, an acid lover (acidophile), flourishes in the hot sulfur springs of Yellowstone National Park. At the opposite end of the pH scale, alkaliphiles occupy carbonate-laden soda lakes around the world. Halophiles are specialized for life in saturated salt lakes and salt evaporation ponds. Others, the barophiles (pressure lovers), colonize the floor of the deepest reaches of the ocean. In 1996, Japanese scientists used a small unmanned submersible to retrieve bottom mud from the Challenger Deep of the Mariana Trench, which at 35,750 feet is the lowest point of the world's oceans. In the samples they discovered hundreds of species of bacteria, archaeans, and fungi. Transferred to the laboratory, some of the bacteria were able to grow at the pressure found in the Challenger Deep, which is a thousand times greater than that near the ocean surface.

The outer reach of physiological resilience of any kind may have been attained by *Deinococcus radiodurans,* a bacterium that can live through radiation so intense the glass of a Pyrex beaker holding them is cooked to a discolored and fragile state. A human being exposed to 1,000 rads of radiation energy, a dose delivered in the atomic explosions at Hiroshima and Nagasaki, dies within one or two weeks. At 1,000 times this amount, 1 million rads, the growth of the *Deinococcus* is slowed, but all the bacteria still survive. At 1.75 million rads, 37 percent make it through, and even at 3 million rads a very small number still endure. The secret of this superbug is its extraordinary ability to repair broken DNA. All organisms have an enzyme that can replace chromosome parts that have been shorn off, whether by radiation, chemical insult, or accident. The more conventional bacterium *Escherichia coli,* a dominant inhabitant of the human gut, can repair two or three breaks at one time. The superbug can manage five hundred breaks. The special molecular techniques it uses remain unknown.

Deinococcus radiodurans and its close relatives are not just extremophiles but ultimate generalists and world travelers, having been found, for example, in llama feces, Antarctic rocks, the tissue of Atlantic haddock, and a can of ground pork and beef irradiated by scientists in Oregon. They join a select group, also including cyanobacteria of the genus *Chroococcidiopsis,* that thrive where very few other organisms venture. They are Earth's outcast nomads, looking for life in all the worst places.

By virtue of their marginality, the superbugs are also candidates for space travel. Microbiologists have begun to ask whether the hardiest among them might drift away from Earth, propelled by stratospheric winds into the void, eventually to settle alive on Mars. Conversely, indigenous microbes from Mars (or beyond) might have colonized Earth. Such is the theory of the origin of life called panspermia, once ridiculed but now an undeniable possibility.

The superbugs have also given a new shot of hope to exobiologists, scientists who look for evidences of life on other worlds. Another stimulus is the newly revealed existence of SLIMEs (subsurface lithoautotrophic microbial ecosystems), unique assemblages of bacteria and fungi that occupy pores in the interlocking mineral grains of igneous rock beneath Earth's surface. Thriving to a depth of up to two miles or more, they obtain their energy from inorganic chemicals. Because they do not require organic particles that filter down from conventional plants and animals whose ultimate energy is from sunlight, the SLIMEs are wholly independent of life on the surface. Consequently, even if all of life as we know it were somehow extinguished, these microscopic troglodytes would carry on. Given enough time, a billion years perhaps, they would likely evolve new forms able to colonize the surface and resynthesize the precatastrophe world run by photosynthesis.

The major significance of the SLIMEs for exobiology is the heightened possibility they suggest of life on other planets and Mars in particular. SLIMEs, or their extraterrestrial equivalent, might live deep within the red planet. During its early, aqueous period Mars had rivers, lakes, and perhaps time to evolve its own surface organisms. According to one recent estimate, there was enough water to cover the entire Martian surface to a depth of five hundred meters. Some, perhaps most, of the water may still exist in permafrost, surface ice covered by the dust we now see from our landers—or, far below the surface, in liquid form. How far below? Physicists believe there is enough heat inside Mars to liquefy water. It comes from a combination of decaying radioactive minerals, some gravitational heat remaining from the original assembly of the planet out of smaller cosmic fragments, and gravitational energy from the sinking of heavier elements and rise of lighter ones. A recent model of the combined effects suggests that the temperature of Mars increases with depth in the upper crustal layers at a rate of 6°F per mile. As a consequence, water could be liquid at eighteen miles beneath the surface. But some water may well up occasionally from the aquifers. In 2000, high-resolution scans by an orbiting satellite revealed the presence of gullies that may have been cut by running streams in the last few centuries or even decades. If Martian life did arise on the planet, or arrived in space particles from Earth, it must include extremophiles, some of which are (or were) ecologically independent single-celled organisms able to persist in or beneath the permafrost.

An equal contender for extraterrestrial life in the solar system is Europa, the second moon out (after Io) of Jupiter. Europa is ice-covered, and long cracks and filled-in meteorite craters on its surface suggest there is an ocean of brine or slurried ice beneath the surface. The evidence is consistent with the likelihood of persistent interior heat in Europa caused by its gravitational tug of war with nearby Jupiter, Io, and Callisto. The main ice crust may be six miles thick, but crisscrossed with far thinner regions on top of upwelling liquid water, thin enough in fact to create slabs that move like icebergs. Do SLIME-like autotrophs float and swim in the Europan Ocean beneath? To planetary scientists and biologists the odds appear good enough to have a look, and practical enough to test—if we can soft-land probes on the upwelling surface cracks and drill through the ice skims that cover them. A second, although less promising, candidate is Callisto, the most distant of

Jupiter's larger moons, which may have an ice crust about sixty miles thick and an underlying salt ocean up to twelve miles deep.

On Earth, the closest approach to the putative oceans of Europa and Callisto is Antarctica's Lake Vostok. About the size of Lake Ontario, with depths exceeding 1,500 feet, Vostok is located under two miles of the East Antarctic Ice Sheet in the remotest part of the continent. It is at least one million years old, wholly dark, under immense pressure, and fully isolated from other ecosystems. If any environment on Earth is sterile, it should be Lake Vostok. Yet this hidden world contains organisms. Scientists have recently drilled through the glacial ice to the six-hundred-foot bottom layer adjacent to the lake. The lowest core samples contained a sparse diversity of bacteria and fungi almost certainly derived from the underlying water. The drill will not be pushed on down into the liquid water. To do so would contaminate one of the last remaining pristine habitats on Earth. The Vostok operation, while telling us very little as yet about the possibility of extraterrestrial life, is a precursor of similar probes likely to be conducted during this century on Mars and the Jovian moons Europa and Callisto.

Suppose that autotrophs parallel to those on Earth originated without benefit of sunlight. Could they have also given rise in the stygian darkness to animals of some kind? The image leaps to mind of crustaceanlike species filtering the microbes and larger, fishlike animals hunting the crustaceoids. A recent discovery on planet Earth suggests that such independent evolution of complex life forms can occur. Romania's Movile Cave was sealed off from the outside more than 5.5 million years ago. During that time it evidently received oxygen through minute cracks in the overlying rocks, but no organic material from the sunlight-driven flora and fauna in the world above. Although the peculiar life forms of most caves around the world draw at least part of their energy from the outside, this is evidently not the case for the Movile Cave and may never have been. The energy base is the autotrophic bacteria, which metabolize hydrogen sulfide from the rocks. Feeding on them and each other are no fewer than forty-eight species of animals, of which thirty-three proved new to science when the cave was explored. The microbe grazers, equivalent to plant eaters on the outside, include pill bugs, springtails, millipedes, and bristletails. Among the carnivores that hunt the microbe grazers are pseudoscorpions, centipedes, and spiders. These more complex organisms are descended from ancestors that entered before the cave was sealed. A second example of an independent stygian system, although not entirely closed to the outside, is Cueva de Villa Luz (Cave of the Lighted House), on the edge of the Chiapas highlands in Tabasco, southern Mexico. Here too the energy base is metabolism by the autotrophic bacteria. Forming layers over the inner cave walls, they subsist on hydrogen sulfide and support a multifarious swarm of small animals.

Studies of the distribution of life have revealed several fundamental patterns in the way species proliferate and are fitted together in Earth's far-flung ecosystems. The first, the most elementary, is that bacteria and archaeans occur everywhere there is life of any kind, whether on the surface or deep beneath it. The second is that, if there is even the smallest space through which to wriggle or swim, tiny protists and invertebrates invade and proceed to prey on the microbes and one another. The third principle is that the more space available, up to and including the largest ecosystems such as grasslands and oceans, the larger are the largest animals living in them. And finally, the greatest diversity of life, as measured by the number of species, occurs in habitats with the most year-round solar energy, the widest exposure of ice-free terrain, the most varied terrain, and the greatest climatic stability across long stretches of time. Thus the equatorial rainforests of the Asian, African, and South American continents possess by far the largest number of plant and animal species.

Regardless of its magnitude, biodiversity (short for biological diversity) is everywhere organized into three levels. At the top are the ecosystems, such as rainforests, coral reefs, and lakes. Next are the species, composed of the organisms in the ecosystems, from algae and swallowtail butterflies to moray eels and people. At the bottom are the variety of genes making up the heredity of individuals that compose each of the species.

Every species is bound to its community in the unique manner by which it variously consumes, is consumed, competes, and cooperates with other species. It also indirectly affects the community in the way it alters the soil, water, and air. The ecologist sees the whole as a network of energy and material continuously flowing into the community from the surrounding physical environment, and back out, and then on round to create the perpetual ecosystem cycles on which our own existence depends.

It is easy to visualize an ecosystem, especially if it is as physically discrete as, say, a marsh or an alpine meadow. But does its dynamical network of organisms, materials, and energy link it to other ecosystems? In 1972 the British inventor and scientist James E. Lovelock said that, in fact, it is tied to the entire biosphere, which can be thought of as a kind of superorganism that surrounds the planet. This singular entity he called Gaia, after Gaea, or Ge, a vaguely personal goddess of early Greece, giver of dreams, divine personification of Earth, and object of the cult of Earth, as well as mother of the seas, the mountains, and the twelve Titans—in other words, *big*. There is considerable merit in looking at life in this grand holistic manner. Alone among the solar planets, Earth's physical environment is held by its organisms in a delicate equilibrium utterly different from what would be the case in their absence. There is plenty of evidence that even some individual species have a measurable global impact. In the most notable example, the oceanic phytoplankton, composed of microscopic, photosynthesizing bacteria, archaeans, and algae, is a major player in the control of the world climate. Dimethylsulfide generated by the algae alone is believed to be an important factor in the regulation of cloud formation.

The concept of the biosphere as Gaia has two versions: strong and weak. The strong version holds that the biosphere is a true superorganism, with each of the species in it optimized to stabilize the environment and benefit from balance in the entire system, like cells of the body or workers of an ant colony. This is a lovely metaphor, with a kernel of truth, providing the idea of superorganism is broadened enough. The strong version, however, is generally rejected by biologists, including Lovelock himself, as a working principle. The weak version, on the other hand, which holds that some species exercise widespread and even global influence, is well substantiated. Its acceptance has stimulated important new programs of research.

Looking at the totality of life, the POET asks, Who are Gaia's children?

The ECOLOGIST responds, They are the species. We must know the role each one plays in the whole in order to manage Earth wisely.

The SYSTEMATIST adds, Then let's get started. How many species exist? Where are they in the world? Who are their genetic kin?

Systematists, the biologists who specialize in classification, favor the species as the unit by which to measure biodiversity. They build on the system of classification invented in the mid-1700s by the Swedish naturalist Carolus Linnaeus. In the Linnaean system each species is given a two-part Latinized name such as *Canis lupus,* for the gray wolf, with *lupus* being the species and *Canis* the genus of wolves and dogs. Similarly, all of humanity composes the species *Homo sapiens.* Today there is only one member of our very distinctive genus, but as recently as 27,000 years ago there was also *Homo neanderthalensis,* the Neanderthal people who preceded *Homo sapiens* in glacier-bound Europe.

The species is the base of the entire Linnaean system and the unit by which biologists traditionally visualize the span of life. The higher categories from genus to domain are simply the means by which the degrees of similarity are subjectively assayed and roughly described. When we say *Homo neanderthalensis,* we mean a species close to *Homo sapiens;* when we say *Australopithecus africanus,* to designate one of the ancestral man-apes, we mean a creature different enough from the species of *Homo* to be placed in another genus, *Australopithecus*. And when we assert that all three of the species composing two genera are hominids, we mean they are close enough to one another to be classified as members of the same family, the Hominidae. The closest living relations of the Hominidae are the common chimpanzee, *Pan troglodytes*, and the pygmy chimpanzee, or bonobo, *Pan paniscus*. They are similar enough to each other, and share sufficiently close common ancestry, to be put in the same genus, *Pan*. And both are different enough from the hominids, with distant enough common ancestry, to constitute not only a distinct genus but a separate family, the Pongidae. The Pongidae also includes a second genus for the orangutan and a third for the two species of gorillas.

And thus in visualizing life we travel nomenclaturally outward through the gossamer pavilions of Earth's biodiversity. The principles of higher classification are very easy to grasp, once you get used to the Latinized names. The Linnaean system builds up hierarchically to the higher categories of biodiversity by the same basic principles used to organize ground combat troops, proceeding from squads to platoons to companies to divisions to corps to armies. Returning to the gray wolf, its genus *Canis*, the common dogs and wolves, are placed into the family Canidae with other genera that hold the species of coyotes and foxes. Families are grouped into orders; the order Carnivora are all the canids plus the families respectively of bears, cats, weasels, raccoons, and hyenas. Orders are clustered into classes, with the class Mammalia composed of the carnivores and all other mammals, and classes are clustered into phyla, in this particular progression the phylum Chordata, which includes mammals and all other vertebrates as well as the vertebra-less lancelets and sea squirts. Thence phyla into kingdoms (Bacteria, Archaea, Protista, Fungi, Animalia, Plantae); and finally, at the summit, encompassing everything, there are the three great domains of life on Earth, the Bacteria, the Archaea, and the Eukarya, the last comprising the protistans (also called protozoans), fungi, animals, and plants.

But always, the real units that can be seen and counted as corporeal objects are the species. Like troops in the field, they are present and waiting to be counted, regardless of how we arbitrarily group and name them. How many species are there in the world? Somewhere between 1.5 and 1.8 million have been discovered and given a formal scientific name. No one has yet made an exact count from the taxonomic literature published over the past 250 years. We know this much, however: the roster, whatever its length, is but a mere beginning. Estimates of the true number of living species range, according to the method used, from 3.6 million to 100 million or more. The median of the estimates is a little over 10 million, but few experts would risk their reputations by insisting on this figure or any other, even to the nearest million.

The truth is that we have only begun to explore life on Earth. How little we know is epitomized by bacteria of the genus *Prochlorococcus,* arguably the most abundant organisms on the planet and responsible for a large part of the organic production of the ocean—yet unknown to science until 1988. *Prochlorococcus* cells float passively in open water at 70,000 to 200,000 per milliliter, multiplying with energy captured from sunlight. Their extremely small size is what makes them so elusive. They belong to a special group called picoplankton, forms even smaller than conventional bacteria and barely visible even at the highest optical magnification.

The blue ocean teems with other novel and little-known bacteria, archaeans, and protozoans. When researchers began to focus on them in the 1990s, they discovered that these organisms are vastly more abundant and diverse than anyone had previously imagined. Much of this miniature world exists in and around previously unseen dark matter, composed of wispy aggregates of colloids, cell fragments, and polymers that range in diameter from billionths to hundredths of a meter. Some of the material contains "hot spots" of nutrients that attract scavenger bacteria and their tiny bacterial and protozoan predators. The ocean we peer into, seemingly clear with only an occasional fish and invertebrate passing beneath, is not the ocean we thought. The visible organisms are just the tip of a vast biomass pyramid.

Among the multicellular organisms of Earth in all environments, the smallest species are also the least known. Of the fungi, which are nearly as ubiquitous as the microbes, 69,000 species have been identified and named, but as many as 1.6 million are thought to exist. Of the nematode worms, making up four of every five animals on Earth and the most widely distributed, 15,000 species are known, but millions more may await discovery.

During the molecular revolution in biology, which spanned the second half of the twentieth century, systematics was judged to be a largely outdated discipline. It was pushed aside and kept on minimal rations. Now the renewal of the Linnaean enterprise is seen as high adventure; systematics has returned to the center of the action in biology. The reasons for the renaissance are multiple. Molecular biology has provided systematics the tools to speed the discovery of microscopic organisms. New techniques are now available in genetics and mathematical tree theory to trace the evolution of life in a swift and convincing manner. All this has happened just in time. The global environmental crisis gives urgency to the full and exact mapping of all biological diversity.

One of the open frontiers in biodiversity exploration is the floor of the ocean, which from surf to abyss covers 70 percent of Earth's surface. All of the thirty-six known animal phyla, the highest-ranking and most inclusive groups in the taxonomic hierarchy, occur there, as opposed to only ten on the land. Among the most familiar are the Arthropoda, or the insects, crustaceans, spiders, and their sundry relations; and the Mollusca, comprising the snails, mussels, and octopuses. Amazingly, two marine phyla have been discovered during the past thirty years: the Loricifera, miniature bullet-shaped organisms with a girdlelike band around their middle, described for the first time in 1983; and the Cycliophora, plump symbiotic forms that attach themselves to the mouths of lobsters and filter out food particles left over from their hosts' meals, described in 1996. Swarming around the loriciferans and cycliophorans, and deep into the soil of shallow marine waters, are other Alice-in-Wonderland creatures, the meiofauna, most of them barely visible to the naked eye. The strange creatures include gastrotrichs, gnathostomulids, kinorhynchs, tardigrades, chaetognaths, placozoans, and orthonectids, along with nematodes and worm-shaped ciliate protozoans. They can be found in buckets of sand drawn from the intertidal surf and offshore shallow water around the world. So, for those seeking a new form of recreation, plan a day at the nearest beach. Take an umbrella, bucket, trowel, microscope, and illustrated textbook on invertebrate zoology. Don't build sand castles but explore, and as you enjoy this watery microcosm keep in mind what the great nineteenth-century physicist Michael Faraday correctly said, that nothing in this world is too wonderful to be true.

Even the most familiar small organisms are less studied than might be guessed. About ten thousand species of ants are known and named, but that number may double when tropical regions are more fully explored. While recently conducting a study of *Pheidole,* one of the world's two largest ant genera, I uncovered 341 new species, more than doubling the number in the genus and increasing the entire known fauna of ants in the Western Hemisphere by 10 percent. As my monograph went to press in 2001, additional new species were still pouring in, mostly from fellow entomologists collecting in the tropics.

You will recognize this frequent image in popular entertainment: a scientist discovers a new species of animal or plant (perhaps after an arduous journey up a tributary of the Orinoco). His team at base camp celebrates, opening a bottle of champagne, and radios the news to the home institution. The truth, I assure you, is almost always different. The small number of scientists expert in the classification of each of the most diverse groups, from bacteria to fungi and insects, are inundated with new species almost to the breaking point. Working mostly alone, they try desperately to keep their collections in order while eking out enough time to publish accounts of a small fraction of the novelties sent to them for identification.

Even the flowering plants, traditionally a favorite of field biologists, retain large pockets of unexamined diversity. About 272,000 species have been described worldwide, but the true number is likely to be 300,000 or more. Each year about 2,000 new species are added to the world list published in botany's standard reference work, the *Index Kewensis*. Even the relatively well-curried United States and Canada continue to yield about 60 new species annually. Some experts believe that as much as 5 percent of the North American flora await discovery, including 300 or more species and races in the biologically rich state of California alone. The novelties are usually rare but not necessarily shy and inconspicuous. Some, like the recently described Shasta snow-wreath *(Neviusia cliftonii),* are flamboyant enough to serve as ornamentals. Many grow in plain sight. A member of the lily family, *Calochortus tiburonensis,* first described in 1972, grows just ten miles from downtown San Francisco. In 1982, a twenty-one-year-old amateur collector, James Morefield, discovered the brand-new leather flower, *Clematis morefieldii,* on the outskirts of Huntsville, Alabama.

Ever deeper rounds of zoological exploration, driven by a sense of urgency over vanishing environments, have revealed surprising numbers of new vertebrates, many of which are placed on the endangered list as soon as they are discovered. The global number of amphibian species, including frogs, toads, salamanders, and the less familiar tropical caecilians, grew between 1985 and 2001 by one third, from 4,003 to 5,282. There can be little doubt that in time it will pass 6,000.

The discovery of new mammals has also continued at a rapid pace. Collectors, by journeying to remote tropical regions and concentrating on small elusive forms such as tenrecs and shrews, have increased the global number in the last two decades from about 4,000 to 5,000. The record for rapid discovery during the past half-century was set by James L. Patton in July 1996. With just three weeks' effort in the central Andes of Colombia, he discovered 6 new species—four mice, a shrew, and a marsupial. Even primates, including apes, monkeys, and lemurs, the most sought of all mammals in the field, are yielding novelties. In the 1990s alone Russell Mittermeier and his colleagues managed to add 9 new species to the 275 previously known. Mittermeier, whose searches take him to tropical forests around the world, estimates that at least another hundred species of primates await discovery.

New land mammals of large size are a rarity, but even a few of them continue to turn up. Perhaps the most surprising find in recent memory was the discovery during the mid-1990s of not one but four big animals in the Annamite Mountains between Vietnam and Laos. Included are a striped hare; a seventy-five-pound barking deer, or giant muntjac; and a smaller, thirty-five-pound barking deer. But most astonishing is the two-hundred-pound cowlike animal called saola, or "spindlehorn," by the local people and Vu Quang bovid by zoologists. It was the first land vertebrate of this size to be discovered for more than fifty years. The saola is not closely related to any other known ungulate mammal. It has been placed in a genus of its own, *Pseudoryx,* meaning false oryx, in reference to its superficial resemblance to the true oryx, a large African antelope. Only a few hundred saola are thought to exist. Their numbers are probably dwindling fast from native hunting and the clearing of the forests in which they live. No scientist has yet seen one in the wild, but in 1998 a photograph was captured by a pressure-released trap camera. And for a short time, before she died, a female brought in by Hmong hunters was kept in the zoo at Lak Xao, Laos.

For centuries, birds have been the most pursued and best known of all animals, but here again new species are still coming to light at a steady pace. From 1920 to 1934, the golden age of ornithological field research, an average of about ten subsequently authenticated species were described each year. The number dropped to between two and three and remained steady thereafter into the 1990s. By the end of the century, approximately ten thousand valid species were securely established in the world register. Then, an unexpected revolution in field studies opened the census to a flood of new candidate species. Experts had come to recognize the possible existence of large numbers of sibling species—populations closely resembling one another in anatomical traits traditionally used in taxonomy, such as size, plumage, and bill shape, yet differing strongly in other, equally important traits discoverable only in the field, such as habitat preference and mating call. The fundamental criterion used to separate species of birds, as well as most other kinds of animals, is that provided by the biological species concept: populations belong to different species if they are incapable of interbreeding freely under natural conditions. As field studies have increased in sophistication, more such genetically isolated populations have come to light. Old species recently subdivided into multiple species include the familiar *Phylloscopus,* leaf warblers, of Europe and Asia and, more controversially, the crossbills of North America. An important new analytic method is song playback, in which ornithologists record the songs of one population and play them in the presence of another population. If the birds show little interest in each other's songs, they can be reasonably assumed to represent different species, because they would presumably not interbreed if they met in nature. The playback method makes possible for the first time the evaluation not only of populations occupying the same range but also those living apart and classified as geographic races, or subspecies. It is not out of the question that the number of validated living bird species will eventually double, to twenty thousand.

More than half the plant and animal species of the world are believed to occur in the tropical rainforests. From these natural greenhouses, which occupy the opposite end of the biodiversity scale from the McMurdo Dry Valleys, many world records of biodiversity have been reported: 425 kinds of trees in a single hectare (2.5 acres) of Brazil's Atlantic Forest, for example, and 1,300 butterfly species from a corner of Peru's Manu National Park. Both numbers are ten times greater than those from comparable sites in Europe and North America. The record for ants is 365 species from 10 hectares (25 acres) in a forest tract of the upper Peruvian Amazon. I have identified 43 species from the canopy of a single *tree* in the same region, approximately equal to the ant fauna of all the British Isles.

These impressive censuses do not exclude a comparable richness of some groups of organisms in other major environments of the world. A single coral head in Indonesia can harbor hundreds of species of crustaceans, polychaete worms, and other invertebrates, plus a fish or two. Twenty-eight kinds of vines and herbaceous plants have been found growing on a giant *Podocarpus* yellowwood conifer in the temperate rainforest of New Zealand, setting the world record for vascular epiphytes on a single tree. As many as two hundred species of mites, diminutive spiderlike creatures, teem in a single square meter of some hardwood forests of North America. In the same spot a gram of soil—a pinch held between thumb and forefinger—contains thousands of species of bacteria. A few are actively multiplying, but most are dormant, each awaiting the special combination of nutrients, moisture, aridity, and temperature to which its particular strain is adapted.

You do not have to visit distant places, or even rise from your seat, to experience the luxuriance of biodiversity. You yourself are a rainforest of a kind. There is a good chance that tiny spiderlike mites build nests at the base of your eyelashes. Fungal spores and hyphae on your toenails await the right conditions to sprout a Lilliputian forest. The vast majority of the cells in your body are not your own; they belong to bacterial and other microorganismic species. More than four hundred such microbial species make their home in your mouth. But rest easy: the bulk of protoplasm you carry around is still human, because microbial cells are so small. Every time you scuff earth or splash mud puddles with your shoes, bacteria, and who knows what else, that are still unknown to science settle on them.

Such is the biospheric membrane that covers Earth, and you and me. It is the miracle we have been given. And our tragedy, because a large part of it is being lost forever before we learn what it is and the best means by which it can be savored and used.

CHAPTER 2

———o———

THE BOTTLENECK

The twentieth century was a time of exponential scientific and technical advance, the freeing of the arts by an exuberant modernism, and the spread of democracy and human rights throughout the world. It was also a dark and savage age of world wars, genocide, and totalitarian ideologies that came dangerously close to global domination. While preoccupied with all this tumult, humanity managed collaterally to decimate the natural environment and draw down the nonrenewable resources of the planet with cheerful abandon. We thereby accelerated the erasure of entire ecosystems and the extinction of thousands of million-year-old species. If Earth's ability to support our growth is finite—and it is—we were mostly too busy to notice.

As a new century begins, we have begun to awaken from this delirium. Now, increasingly postideological in temper, we may be ready to settle down before we wreck the planet. It is time to sort out Earth and calculate what it will take to provide a satisfying and sustainable life for everyone into the indefinite future. The question of the century is: How best can we shift to a culture of permanence, both for ourselves and for the biosphere that sustains us?

The bottom line is different from that generally assumed by our leading economists and public philosophers. They have mostly ignored the numbers that count. Consider that with the global population past six billion and on its way to eight billion or more by mid-century, per-capita fresh water and arable land are descending to levels resource experts agree are risky. The ecological footprint—the average amount of productive land and shallow sea appropriated by each person in bits and pieces from around the world for food, water, housing, energy, transportation, commerce, and waste absorption—is about one hectare (2.5 acres) in developing nations but about 9.6 hectares (24 acres) in the United States. The footprint for the total human population is 2.1 hectares (5.2 acres). For every person in the world to reach present U.S. levels of consumption with existing technology would require four more planet Earths. The five billion people of the developing countries may never wish to attain this level of profligacy. But in trying to achieve at least a decent standard of living, they have joined the industrial world in erasing the last of the natural environments. At the same time *Homo sapiens* has become a geophysical force, the first species in the history of the planet to attain that dubious distinction. We have driven atmospheric carbon dioxide to the highest levels in at least two hundred thousand years, unbalanced the nitrogen cycle, and contributed to a global warming that will ultimately be bad news everywhere.

In short, we have entered the Century of the Environment, in which the immediate future is usefully conceived as a bottleneck. Science and technology, combined with a lack of self-understanding and a Paleolithic obstinacy, brought us to where we are today. Now science and technology, combined with foresight and moral courage, must see us through the bottleneck and out.

"Wait! Hold on there just one minute!"

That is the voice of the cornucopian economist. Let us listen to him carefully. You can read him in the pages of *The Economist, The Wall Street Journal,* and myriad white papers prepared for the

Competitive Enterprise Institute and other politically conservative think tanks. I will use these sources to synthesize his position, as honestly as I can, recognizing the dangers of stereotyping. He will meet an ecologist, in order to have a congenial dialogue. Congenial, because it is too late in the day for combat and debating points. Let us make the honorable assumption that economist and ecologist have as a common goal the preservation of life on this beautiful planet.

The economist is focused on production and consumption. These are what the world wants and needs, he says. He is right, of course. Every species lives on production and consumption. The tree finds and consumes nutrients and sunlight; the leopard finds and consumes the deer. And the farmer clears both away to find space and raise corn—for consumption. The economist's thinking is based on precise models of rational choice and near-horizon time lines. His parameters are the gross domestic product, trade balance, and competitive index. He sits on corporate boards, travels to Washington, occasionally appears on television talk shows. The planet, he insists, is perpetually fruitful and still underutilized.

The ecologist has a different worldview. He is focused on unsustainable crop yields, overdrawn aquifers, and threatened ecosystems. His voice is also heard, albeit faintly, in high government and corporate circles. He sits on nonprofit foundation boards, writes for *Scientific American,* and is sometimes called to Washington. The planet, he insists, is exhausted and in trouble.

THE ECONOMIST

"Ease up. In spite of two centuries of doomsaying, humanity is enjoying unprecedented prosperity. There are environmental problems, certainly, but they can be solved. Think of them as the detritus of progress, to be cleared away. The global economic picture is favorable. The gross national products of the industrial countries continue to rise. Despite their recessions, the Asian tigers are catching up with North America and Europe. Around the world, manufacture and the service economy are growing geometrically. Since 1950 per-capita income and meat production have risen continuously. Even though the world population has increased at an explosive 1.8 percent each year during the same period, cereal production, the source of more than half the food calories of the poorer nations and the traditional proxy of worldwide crop yield, has more than kept pace, rising from 275 kilograms per head in the early 1950s to 370 kilograms by the 1980s. The forests of the developed countries are now regenerating as fast as they are being cleared, or nearly so. And while fibers are also declining steeply in most of the rest of the world—a serious problem, I grant—no global scarcities are expected in the foreseeable future. Agriforestry has been summoned to the rescue: more than 20 percent of industrial wood fiber now comes from tree plantations.

"Social progress is running parallel to economic growth. Literacy rates are climbing, and with them the liberation and empowerment of women. Democracy, the gold standard of governance, is spreading country by country. The communication revolution powered by the computer and the Internet has accelerated the globalization of trade and the evolution of a more irenic international culture.

"For two centuries the specter of Malthus troubled the dreams of futurists. By rising exponentially, the doomsayers claimed, population must outstrip the limited resources of the world and bring about famine, chaos, and war. On occasion this scenario did unfold locally. But that has been more the result of political mismanagement than Malthusian mathematics. Human ingenuity has always found a way to accommodate rising populations and allow most to prosper. The green revolution, which dramatically raised crop yields in the developing countries, is the outstanding example. It can be repeated with new technology. Why should we doubt that human entrepreneurship can keep us on an upward-turning curve?

"Genius and effort have transformed the environment to the benefit of human life. We have turned a wild and inhospitable world into a garden. Human dominance is Earth's destiny. The harmful perturbations we have caused can be moderated and reversed as we go along."

THE ENVIRONMENTALIST

"Yes, it's true that the human condition has improved dramatically in many ways. But you've painted only half the picture, and with all due respect the logic it uses is just plain dangerous. As your worldview implies, humanity has learned how to create an economy-driven paradise. Yes again—but only on an infinitely large and malleable planet. It should be obvious to you that Earth is finite and its environment increasingly brittle. No one should look to GNPs and corporate annual reports for a competent projection of the world's long-term economic future. To the information there, if we are to understand the real world, must be added the research reports of natural-resource specialists and ecological economists. They are the experts who seek an accurate balance sheet, one that includes a full accounting of the costs to the planet incurred by economic growth.

"This new breed of analysts argues that we can no longer afford to ignore the dependency of the economy and social progress on the environmental resource base. It is the *content* of economic growth, with natural resources factored in, that counts in the long term, not just the yield in products and currency. A country that levels its forests, drains its aquifers, and washes its topsoil downriver without measuring the cost is a country traveling blind. It faces a shaky economic future. It suffers the same delusion as the one that destroyed the whaling industry. As harvesting and processing techniques were improved, the annual catch of whales rose, and the industry flourished. But the whale populations declined in equal measure until they were depleted. Several species, including the blue whale, the largest animal species in the history of Earth, came close to extinction. Whereupon most whaling was called to a halt. Extend that argument to falling ground water, drying rivers, and shrinking per-capita arable land, and you get the picture.

"Suppose that the conventionally measured global economic output, now at about $31 trillion, were to expand at a healthy 3 percent annually. By 2050 it would in theory reach $138 trillion. With only a small leveling adjustment of this income, the entire world population would be prosperous by current standards. Utopia at last, it would seem! What is the flaw in the argument? It is the environment crumbling beneath us. If natural resources, particularly fresh water and arable land, continue to diminish at their present per-capita rate, the economic boom will lose steam, in the course of which—and this worries me even if it doesn't worry you—the effort to enlarge productive land will wipe out a large part of the world's fauna and flora.

"The appropriation of productive land—the ecological footprint—is already too large for the planet to sustain, and it's growing larger. A recent study building on this concept estimated that the human population exceeded Earth's sustainable capacity around the year 1978. By 2000 it had overshot by 1.4 times that capacity. If 12 percent of land were now to be set aside in order to protect the natural environment, as recommended in the 1987 Brundtland Report, Earth's sustainable capacity will have been exceeded still earlier, around 1972. In short, Earth has lost its ability to regenerate—unless global consumption is reduced, or global production is increased, or both."

By dramatizing these two polar views of the economic future, I don't wish to imply the existence of two cultures with distinct ethos. All who care about both the economy and environment, and that includes the vast majority, are members of the same culture. The gaze of our two debaters is fixed on different points in the space-time scale in which we all dwell. They differ in the factors they take into account in forecasting the state of the world, how far they look into the future, and how much they care about nonhuman life. Most economists today, and all but the most politically conservative of their public interpreters, recognize very well that the world has limits and the human population cannot afford to grow much larger. They know that humanity is destroying biodiversity. They just don't like to spend a lot of time thinking about it.

The environmentalist view is fortunately spreading. Perhaps the time has come to cease calling it the "environmentalist" view, as though it were a lobbying effort outside the mainstream of human activity, and to start calling it the real-world view. In a realistically reported and managed economy, balanced accounting will be routine. The conventional gross national product (GNP) will be replaced by the more comprehensive genuine progress indicator (GPI), which includes estimates of environmental costs of economic activity. Already, a growing number of economists, scientists, political leaders, and others have endorsed precisely this change.

What, then, are essential facts about population and environment? From existing databases we can answer that question and visualize more clearly the bottleneck through which humanity and the rest of life are now passing.

On or about October 12, 1999, the world population reached 6 billion. It has continued to climb at an annual rate of 1.4 percent, adding 200,000 people each day or the equivalent of the population of a large city each week. The rate, although beginning to slow, is still basically exponential: the more people, the faster the growth, thence still more people sooner and an even faster growth, and so on upward toward astronomical numbers unless the trend is reversed and growth rate is reduced to zero or less. This exponentiation means that people born in 1950 were the first to see the human population double in their lifetime, from 2.5 billion to over 6 billion now. During the twentieth century more people were added to the world than in all of previous human history. In 1800 there had been about 1 billion; and in 1900, still only 1.6 billion.

The pattern of human population growth in the twentieth century was more bacterial than primate. When *Homo sapiens* passed the six billion mark we had already exceeded by as much as a hundred times the biomass of any large animal species that ever existed on the land. We and the rest of life cannot afford another hundred years like that.

By the end of the century some relief was in sight. In most parts of the world—North and South America, Europe, Australia, and most of Asia—people had begun gingerly to tap the brake pedal. The worldwide average number of children per woman fell from 4.3 in 1960 to 2.6 in 2000. The number required to attain zero population growth—that is, the number that balances the birth and death rates and holds the standing population size constant—is 2.1 (the extra one-tenth compensates for infant and child mortality). When the number of children per woman stays above 2.1 even slightly, the population still expands exponentially. This means that although the population climbs less and less steeply as the number approaches 2.1, humanity will still, in theory, eventually come to weigh as much as the Earth and, if given enough time, will exceed the mass of the visible universe. This fantasy is a mathematician's way of saying that anything above zero population growth cannot be sustained. If, on the other hand, the average number of children drops below 2.1, the population enters negative exponential growth and starts to decline. To speak of 2.1 in exact terms as the breakpoint is of course an oversimplification. Advances in medicine and public health can lower the breakpoint toward the minimal, perfect number of 2.0 (no infant or childhood deaths), while famine, epidemics, and war, by boosting mortality, can raise it well above 2.1. But worldwide, over an extended period of time, local differences and statistical fluctuations wash one another out and the iron demographic laws grind on. They transmit to us always the same essential message, that to breed in excess is to overload the planet.

By 2000 the replacement rate in all of the countries of Western Europe had dropped below 2.1. The lead was taken by Italy, at 1.2 children per woman (so much for the power of natalist religious doctrine). Thailand also passed the magic number, as well as the nonimmigrant population of the United States.

When a country descends to its zero-population birthrates, or even well below, it does not cease absolute population growth immediately, because the positive growth experienced just before the breakpoint has generated a disproportionate number of young people with most of their fertile years and life ahead of them. As this cohort ages, the proportion of child-bearing people diminishes, the age distribution stabilizes at the zero-population level, the slack is taken up, and population growth ceases. Similarly, when a country dips below the breakpoint, a lag period intervenes before the absolute growth rate goes negative and the population actually declines. Italy and Germany, for example, have entered a period of such true, absolute negative population growth.

The decline in global population growth is attributable to three interlocking social forces: the globalization of an economy driven by science and technology, the consequent implosion of rural populations into cities, and, as a result of globalization and urban implosion, the empowerment of women. The freeing of women socially and economically results in fewer children. Reduced reproduction by female choice can be thought a fortunate, indeed almost miraculous, gift of human nature to future generations. It could have gone the other way: women, more prosperous and less shackled, could have chosen the satisfactions of a larger brood. They did the opposite. They opted for a smaller number of quality children, who can be raised with better health and education, over a larger family. They simultaneously chose better, more secure lives for themselves. The tendency appears to be very widespread, if not universal. Its importance cannot be overstated. Social commentators often remark that humanity is endangered by its own instincts, such as tribalism, aggression, and personal greed.

Demographers of the future will, I believe, point out that on the other hand humanity was saved by this one quirk in the maternal instinct.

The global trend toward smaller families, if it continues, will eventually halt population growth, and afterward reverse it. World population will peak and then start down. What will be the peak, and when will it occur? And how will the environment fare as humanity climbs to the peak? In September 1999 the Population Division of the United Nations Department of Economic and Social Affairs released a spread of projections to the year 2050 based on four possible scenarios of female fertility. If the number of children dropped to two per woman immediately—in other words, beginning in 2000—the world population would be on its way to leveling off by around 2050 to approximately 7.3 billion. This degree of descent has not happened of course and is unlikely to be attained for at least several more decades. Thus, 7.3 billion is improbably low. If, at the other extreme, fertility continues to fall at the current rate, the population will reach 10.7 billion by 2050 and continue steeply upward for a few more decades before peaking. If it holds to the present growth rate, it will reach 14.4 billion by 2050. Finally, if fertility falls more rapidly than the present trend, on its way to global 2.1 and below, the population will reach 8.9 billion by 2050; in this case also it will continue to climb for a while longer, but less steeply so. This final scenario appears to be the most likely of the trends. Very broadly, then, it seems probable that the world population will peak in the late twenty-first century somewhere between 9 and 10 billion. If population control efforts are intensified, the number can be brought closer to 9 than to 10 billion.

Enough slack still exists in the system to justify guarded optimism. Women given a choice and affordable contraceptive methods generally practice birth control. The percentage who do so still varies enormously among countries. Europe and the United States, for example, have topped 70 percent; Thailand and

Colombia are closing on that figure; Indonesia is up to about 50 percent; Bangladesh and Kenya have passed 30 percent; but Pakistan holds with little change at around 10 percent. The stated intention, or at least the acquiescence, of national governments favors a continued rise in the levels of birth control worldwide. By 1996, about 130 countries subsidized family planning services. More than half of all developing countries in particular also had official population policies to accompany their economic and military policies, and more than 90 percent of the rest stated their intention to follow suit. The United States, where the idea is still virtually taboo, remained a stunning exception.

The encouragement of population control by developing countries comes not a moment too soon. The environmental fate of the world lies ultimately in their hands. They now account for virtually all global population growth, and their drive toward higher per-capita consumption will be relentless.

The consequences of their reproductive prowess are multiple and deep. The people of the developing countries are already far younger than those in the industrial countries and destined to become more so. The streets of Lagos, Manaus, Karachi, and other cities in the developing world are a sea of children. To an observer fresh from Europe or North America the crowds give the feel of a gigantic school just let out. In at least sixty-eight of the countries, more than 40 percent of the population is under fifteen years of age. Here are typical examples reported in 1999: Afghanistan, 42.9 percent; Benin, 47.9; Cambodia, 45.4; Ethiopia, 46.0; Grenada, 43.1; Haiti, 42.6; Iraq, 44.1; Libya, 48.3; Nicaragua, 44.0; Pakistan, 41.8; Sudan, 45.4; Syria, 46.1; Zimbabwe, 43.8.

A country poor to start with and composed largely of young children and adolescents is strained to provide even minimal health services and education for its people. Its superabundance of cheap, unskilled labor can be turned to some economic advantage but unfortunately also provides cannon fodder for ethnic strife and war. As the populations continue to explode and water and arable land grow scarcer, the industrial countries will feel their pressure in the form of many more desperate immigrants and the risk of spreading international terrorism. I have come to understand the advice given me many years ago when I argued the case for the natural environment to the president's scientific advisor: your patron is foreign policy.

Stretched to the limit of its capacity, how many people can the planet support? A rough answer is possible, but it is a sliding one contingent on three conditions: how far into the future the planetary support is expected to last, how evenly the resources are to be distributed, and the quality of life most of humanity expects to achieve. Consider food, which economists commonly use as a proxy of carrying capacity. The current world production of grains, which provide most of humanity's calories, is about 2 billion tons annually. That is enough, in theory, to feed 10 billion East Indians, who eat primarily grains and very little meat by Western standards. But the same amount can support only about 2.5 billion Americans, who convert a large part of their grains into livestock and poultry. The ability of India and other developing countries to climb the trophic chain is problematic. If soil erosion and withdrawal of groundwater continue at their present rates until the world population reaches (and hopefully peaks) at 9 to 10 billion, shortages of food seem inevitable. There are two ways to stop short of the wall. Either the industrialized populations move down the food chain to a more vegetarian diet, or the agricultural yield of productive land worldwide is increased by more than 50 percent.

The constraints of the biosphere are fixed. The bottleneck through which we are passing is real. It should be obvious to anyone not in a euphoric delirium that whatever humanity does or does not do, Earth's capacity to support our species is approaching the limit. We already appropriate 40 percent of the planet's organic matter produced by green plants. If everyone agreed to become vegetarian, leaving little or nothing for livestock, the present 1.4 billion hectares of arable land (3.5 billion acres) would support about 10 billion people. If humans utilized as food all of the energy captured by plant photosynthesis on land and sea, some 40 trillion watts, the planet could support about 17 billion people. But long before that ultimate limit was approached, the planet would surely have become a hellish place to exist. There may, of course, be escape hatches. Petroleum reserves might be converted into food, until they are exhausted. Fusion energy could conceivably be used to create light, whose energy would power photosynthesis, ramp up plant growth beyond that dependent on solar energy, and hence create more food. Humanity might even consider becoming someday what the astrobiologists call a type II civilization, and harness all the power of the sun to support human life on Earth and on colonies on and around the other solar planets. (No intelligent life forms in the Milky Way galaxy are likely at this level; otherwise they would probably have been already detected by the search for extraterrestrial intelligence, or SETI, programs.) Surely these are not frontiers we will wish to explore in order simply to continue our reproductive folly.

The epicenter of environmental change, the paradigm of population stress, is the People's Republic of China. By 2000 its population was 1.2 billion, one-fifth of the world total. It is thought likely by demographers to creep up to 1.6 billion by 2030. During 1950–2000 China's people grew by 700 million, more than existed in the entire world at the start of the industrial revolution. The great bulk of this increase is crammed into the basins of the Yangtze and Yellow Rivers, covering an area about equal to that of the eastern United States. Americans, when they started from roughly the same point, found themselves geographically blessed. During their own population explosion, from 2 million at the birth of the republic in 1776 to 270 million in 2000, they were able to spread across a fertile and essentially empty continent. The surplus of people, flowing like a tidal wave westward, filled the Ohio Valley, Great Plains, and finally the valleys of the Pacific

Coast. The Chinese could not flow anywhere. Hemmed in to the west by deserts and mountains, limited to the south by resistance from other civilizations, their agricultural populations simply grew denser on the land their ancestors had farmed for millennia. China became in effect a great overcrowded island, a Jamaica or Haiti writ large.

Highly intelligent and innovative, its people have made the most of it. Today China and the United States are the two leading grain producers of the world. The two countries grow a disproportionate share of the food from which the world population derives most of its calories. But China's huge population is on the verge of consuming more than it can produce. In 1997 a team of scientists, reporting to the U.S. National Intelligence Council (NIC), predicted that China will need to import 175 million tons of grain annually by 2025. Extrapolated to 2030, the annual level is 200 million tons—the entire amount of grain exported annually at the present time. A tick in the parameters of the model could move these figures up or down, but optimism would be a dangerous attitude in planning strategy when the stakes are so high. After 1997 the Chinese in fact instituted a province-level crash program to boost grain level to export capacity. The effort was successful but may be short-lived, a fact the government itself recognizes. It requires cultivation of marginal land, higher per-acre environmental damage, and a more rapid depletion of the country's precious ground water.

According to the NIC report, any slack in China's production may be picked up by the Big Five grain exporters, the United States, Canada, Argentina, Australia, and the European Union. But the exports of these dominant producers, after climbing steeply in the 1960s and 1970s, tapered off to near their present level in 1980. With existing agricultural capacity and technology, this output does not seem likely to increase to any significant degree. The United States and the European Union have already returned to production all of the cropland idled under earlier farm commodity programs. Australia and Canada, largely dependent on dryland farming, are constrained by low rainfall. Argentina has the potential to expand, but due to its small size the surplus it produces is unlikely to exceed ten million tons of grain production per year.

China relies heavily on irrigation with water drawn from its aquifers and great rivers. The greatest impediment is again geographic: two-thirds of China's agriculture is in the north but four-fifths of the water supply is in the south—that is, principally in the Yangtze River Basin. Irrigation and withdrawals for domestic and industrial use have depleted the northern basins, from which flow the waters of the Yellow, Hai, Huai, and Liao Rivers. Added to the Yangtze Basin, these regions produce three-fourths of China's food and support 900 million of its population. Starting in 1972, the Yellow River channel has gone bone dry almost yearly through part of its course in Shandong Province, as far inland as the capital, Jinan, thence down all the way to the sea. In 1997 the river stopped flowing for 130 days, then restarted and stopped again through the year for a record total of 226 dry days. Because Shandong Province normally produces a fifth of China's wheat and a seventh of its corn, the failure of the Yellow River is of no little consequence. The crop losses in 1997 alone reached $1.7 billion.

Meanwhile, the groundwater of the northern plains has dropped precipitously, reaching an average 1.5 meters (5 feet) per year by the mid-1990s. Between 1965 and 1995 the water table fell 37 meters (121 feet) beneath Beijing itself.

Faced with chronic water shortages in the Yellow River Basin, the Chinese government has undertaken the building of the Xiaolangdi Dam, which will be exceeded in size only by the Three Gorges Dam on the Yangtze River. The Xiaolangdi is expected to solve the problems of both periodic flooding and drought. Plans are being laid in addition for the construction of canals to siphon water from the Yangtze, which never grows dry, to the Yellow River and Beijing respectively.

These measures may or may not suffice to maintain Chinese agriculture and economic growth. But they are complicated by formidable side effects. Foremost is silting from the upriver loess plains, which makes the Yellow River the most turbid in the world and threatens to fill the Xiaolangdi Reservoir, according to one study, as soon as thirty years after its completion.

China has maneuvered itself into a position that forces it continually to design and redesign its lowland territories as one gigantic hydraulic system. But this is not the fundamental problem. The fundamental problem is that China has too many people. In addition, its people are admirably industrious and fiercely upwardly mobile. As a result their water requirements, already oppressively high, are rising steeply. By 2030 residential demands alone are projected to increase more than fourfold to 134 billion tons, and industrial demands fivefold to 269 billion tons. The effects will be direct and powerful. Of China's 617 cities, 300 already face water shortages.

The pressure on agriculture is intensified in China by a dilemma shared in varying degrees by every country. As industrialization proceeds, per-capita income rises, and the populace consumes more food. They also migrate up the energy pyramid to meat and dairy products. Because fewer calories per kilogram of grain are obtained when first passed through poultry and livestock instead of being eaten directly, per-capita grain consumption rises still more. All the while the available water supply remains static or nearly so. In an open market, the agricultural use of water is outcompeted by industrial use. A thousand tons of fresh water yields a ton of wheat, worth $200, but the same amount of water in industry yields $14,000. As China, already short on water and arable land, grows more prosperous through industrialization and trade, water becomes more expensive. The cost of agriculture rises correspondingly, and, unless the collection of water is subsidized, the price of food also rises. This is in part the rationale for the great dams at Three Gorges and Xiaolangdi, built at enormous public expense.

In theory, an affluent industrialized country does not have to be agriculturally independent. In theory, China can make up its grain shortage by purchasing from the Big Five grain-surplus nations. Unfortunately, its population is too large and the world surplus too restrictive for it to solve its problem without altering the world market. All by itself, China seems destined to drive up the price of grain and make it harder for the poorer developing countries to meet their own needs. At the present time grain prices are falling, but this seems certain to change as the world population soars to 9 billion and beyond.

The problem, resource experts agree, cannot be solved entirely by hydrological engineering. It must include shifts from grain to fruit and vegetables, which are more labor-intensive, giving China a competitive edge. To this can be added strict water conservation measures in industrial and domestic use; the use of sprinkler and drip irrigation in cultivation, as opposed to the traditional and more wasteful methods of flood and furrow irrigation; and private land ownership, with subsidies and price liberalization, to increase conservation incentives for farmers.

Meanwhile, the surtax levied on the environment to support China's growth, although rarely entered on the national balance sheets, is escalating to a ruinous level. Among the most telling indicators is the pollution of water. Here is a measure worth pondering. China has in all 50,000 kilometers of major rivers. Of these, according to the U.N. Food and Agriculture Organization, 80 percent no longer support fish. The Yellow River is dead along much of its course, so fouled with chromium, cadmium, and other toxins from oil refineries, paper mills, and chemical plants as to be unfit for either human consumption or irrigation. Diseases from bacterial and toxic-waste pollution are epidemic.

China can probably feed itself to at least mid-century, but its own data show that it will be skirting the edge of disaster even as it accelerates its life-saving shift to industrialization and mega-hydrological engineering. The extremity of China's condition makes it vulnerable to the wild cards of history. A war, internal political turmoil, extended droughts, or crop disease can kick the economy into a downspin. Its enormous population makes rescue by other countries impracticable.

China deserves close attention, not just as the unsteady giant whose missteps can rock the world, but also because it is so far advanced along the path to which the rest of humanity seems inexorably headed. If China solves its problems, the lessons learned can be applied elsewhere. That includes the United States, whose citizens are working at a furious pace to overpopulate and exhaust their own land and water from sea to shining sea.

Environmentalism is still widely viewed, especially in the United States, as a special-interest lobby. Its proponents, in this blinkered view, flutter their hands over pollution and threatened species, exaggerate their case, and press for industrial restraint and the protection of wild places, even at the cost of economic development and jobs.

Environmentalism is something more central and vastly more important. Its essence has been defined by science in the following way. Earth, unlike the other solar planets, is not in physical equilibrium. It depends on its living shell to create the special conditions on which life is sustainable. The soil, water, and atmosphere of its surface have evolved over hundreds of millions of years to their present condition by the activity of the biosphere, a stupendously complex layer of living creatures whose activities are locked together in precise but tenuous global cycles of energy and transformed organic matter. The biosphere creates our special world anew every day, every minute, and holds it in a unique, shimmering physical disequilibrium. On that disequilibrium the human species is in total thrall. When we alter the biosphere in any direction, we move the environment away from the delicate dance of biology. When we destroy ecosystems and extinguish species, we degrade the greatest heritage this planet has to offer and thereby threaten our own existence.

Humanity did not descend as angelic beings into this world. Nor are we aliens who colonized Earth. We evolved here, one among many species, across millions of years, and exist as one organic miracle linked to others. The natural environment we treat with such unnecessary ignorance and recklessness was our cradle and nursery, our school, and remains our one and only home. To its special conditions we are intimately adapted in every one of the bodily fibers and biochemical transactions that gives us life.

That is the essence of environmentalism. It is the guiding principle of those devoted to the health of the planet. But it is not yet a general worldview, evidently not yet compelling enough to distract many people away from the primal diversions of sport, politics, religion, and private wealth.

The relative indifference to the environment springs, I believe, from deep within human nature. The human brain evidently evolved to commit itself emotionally only to a small piece of geography, a limited band of kinsmen, and two or three generations into the future. To look neither far ahead nor far afield is elemental in a Darwinian sense. We are innately inclined to ignore any distant possibility not yet requiring examination. It is, people say, just good common sense. Why do they think in this short-sighted way? The reason is simple: it is a hard-wired part of our Paleolithic heritage. For hundreds of millennia those who worked for short-term gain within a small circle of relatives and friends lived longer and left more offspring—even when their collective striving caused their chiefdoms and empires to crumble around them. The long view that might have saved their distant descendants required a vision and extended altruism instinctively difficult to marshal.

The great dilemma of environmental reasoning stems from this conflict between short-term and long-term values. To select values for the near future of one's own tribe or country is relatively easy. To select values for the distant future of the whole planet also is relatively easy—in theory at least. To combine the two visions to create a universal environmental ethic is, on the other hand, very difficult. But combine them we must, because a universal environmental ethic is the only guide by which humanity and the rest of life can be safely conducted through the bottleneck into which our species has foolishly blundered.

CHAPTER 3

———○———

NATURE'S LAST STAND

The wealth of the world, if measured by domestic product and per-capita consumption, is rising. But if calculated from the condition of the biosphere, it is falling. The state of the latter, natural economy, as opposed to that of the former, market economy, is measured by the condition of the world's forest, freshwater, and marine ecosystems. When distilled from the databases of the World Bank and United Nations Development and Environment Program as a single Living Planet Index, the result forms a powerful counterweight to the more familiar GNPs and stock market indexes. From 1970 to 1995 the index, as calculated by the World Wide Fund for Nature, fell 30 percent. By the early 1990s its decline had accelerated to 3 percent per year. No leveling trend is yet in sight.

Environmental indices are not popular topics at international economic conferences. In the climate-controlled hotel and meeting rooms the attendees inhabit, the faraway fall of ancient forests and extinction of species are easily deferred as "externalities." Heads of state and finance ministers know they will win few points back home for signing agreements on global conservation.

By and large, religious leaders also lack a record in environmental stewardship of which they can be proud. Even though the fate

of the creation itself is at stake, very few are committed conservationists. Seen from a historical perspective, however, the hesitancy
of the majority is understandable. The sacred texts of the Abrahamic religions contain few instructions about the rest of the living world. The Iron Age scribes who wrote them knew war. They
knew love and compassion. They knew purity of spirit. But they
did not know ecology.

A more realistic view of the human prospect is now in order.
Overpopulation and environmentally ignorant development are
everywhere shrinking natural habitats and biological diversity. In
the real world, governed equally by the market and natural economies, humanity is in a final struggle with the rest of life. If it presses
on, it will win a Cadmean victory, in which first the biosphere
loses, then humanity.

A typical battlefield of this struggle is Hawaii, America's most
deceptively beautiful state. For most residents and visitors, it seems
an unspoiled island paradise. In actuality it is a killing field of biological diversity. When the first Polynesian voyagers put ashore
around A.D. 400, the archipelago was as close to Eden as any land
that ever existed. Its lush forests and fertile valleys contained no
mosquitoes, no ants, no stinging wasps, no venomous snakes or
spiders, and few plants with thorns or poisons. All these infelici-
ties are abundant now, having been introduced by human commerce, both deliberately and accidentally.

Prehuman Hawaii was biologically diverse and unique. It
teemed from shore to mountain peak with at least 125 and as many
as 145 species of birds found nowhere else. Native eagles soared
above thick forests that were home to strange long-legged owls
and a glittering array of painted honeycreepers. On the ground a
species of flightless ibis foraged alongside the moa nalo, goose-
sized flightless birds with jaws vaguely resembling those of tortoises, Hawaii's version of the Mauritian dodo. Almost all of these
endemic forms are now extinct. Only 35 of the original species of
birds still exist, of which 24 are endangered, with a dozen so rare
they may be beyond recovery. A few of the survivors, mostly small

honeycreepers, can still be glimpsed in scattered lowland habitats. The majority, however, cling to life in densely forested upland valleys with high rainfall, as far removed from human presence on the islands as possible. "To see Hawaiian birds," the ornithologist Stuart L. Pimm reported after a series of field studies, "one gets cold, wet and tired."

Hawaii's biodiversity today is still rich. But it is highly synthetic: the vast majority of plants and animals easy to find originated somewhere else. In the immigrant vegetation in and around the resorts and hillside copses live a riot of skylarks, barred and spotted doves, hill robins, mockingbirds, bush warblers, mynahs, strawberry finches, ricebirds, and red-crested cardinals, none of which is native. Like the tourists admiring them, they traveled to Hawaii by boat and aircraft from continents far away. The same species can be seen with monotonous frequency elsewhere in the warm temperate and tropical zones of the world.

Hawaii's plant life is also beautiful, in places extravagantly so. But very few of the species that dominate the lowland vegetation were among those through which the Polynesian colonists first cut their way. Of the 1,935 free-living flowering plant species recognized by botanists today, 902 are alien, and these introduced forms dominate all but the least disturbed habitats. Even the vegetation investing the most natural-appearing habitats of the coastal lowlands and lower mountain slopes is primarily introduced. Hawaii's verdant glens are biogeographic facades populated mostly or entirely by alien species. The leis placed around the necks of arriving visitors are, appropriately, made from immigrant flowers.

Once there were over ten thousand kinds of plants and animals native to Hawaii. Many ranked among the most distinctive and beautiful in the world. They had descended during millions of years from several hundred pioneer species lucky enough to make an unassisted landfall on this most distant of Earth's archipelagoes. That number has now been drastically whittled down. Ancient

Hawaii is a ghost that haunts the hills, and our planet is poorer for its sad retreat.

It all began when the early Polynesians, finding the flight-less birds easy prey, evidently hunted them to extinction. The colonists erased other plant and animal species when they cleared forests and grasslands for agriculture. Captain James Cook, making the European discovery of Hawaii in 1778 in command of HMS *Resolution,* observed that large stretches of the lowlands and interior mountain foothills were covered by plantations of bananas, breadfruit, and sugarcane. In the following two centuries, American and other colonists appropriated both these fields and large stretches of additional terrain to turn sugarcane and pineapple into major export crops. Today barely one-quarter of the land remains untouched, and that is largely confined to the steepest and least accessible redoubts of the mountainous interior. Had Hawaii been flat, like Barbados and the Pacific atolls, for example, virtually nothing would be left.

At first habitat destruction was the principal destroyer of the Hawaiian fauna and flora, but the greatest threat today is from nonnative organisms. The biota of prehistoric Hawaii was relatively small and vulnerable. When the islands were colonized, and especially in the twentieth century when they became the commercial and transportation hub of the Pacific, alien plants, animals, and microbes flooded in from all over other warm temperate and tropical regions of the planet, pushing back and extirpating the native species.

The invasion of Hawaii can be viewed as an abnormal acceleration of the Darwinian process. Before the arrival of humanity, members of perhaps one immigrant species made a successful Pacific crossing every thousand years. Some of the voyagers traveled on air currents in the upper atmosphere. Wings were not an absolute requirement for flight: large numbers of flightless organisms are routinely caught in updrafts and carried by the wind as part of a passive aerial plankton. Many kinds of spiders enter the plankton deliberately. Standing on a leaf or twig, they spin out silk into a passing breeze, letting the threads grow longer and longer until, like swelling balloons, they tug strongly at the body of the spider. The spiders then let go and sail upward. If they catch the right updraft and prevailing wind they can travel long distances before settling to the ground or—fatally—into the water. Some descend voluntarily by reeling in and eating the silk lines. Native spiders, it should come as no surprise, are abundant and diverse on Hawaii.

Less sophisticated travelers are lifted by storm winds and ride them to the islands, or else make the journey as passengers on natural rafts formed by fallen riverside trees and other masses of vegetation swept by floodwater out to sea.

The odds against each individual voyager settling prehuman Hawaii were astronomical. For millions of years many species began the blind journey across the Pacific, but extremely few made a successful landfall. Even then the pioneers faced formidable obstacles. There had to be a niche to fill immediately upon arrival—the right place to live, the right food to eat, potential mates immigrating with them, and few or no predators waiting to gobble them up. Species able to survive and breed were then candidates for evolutionary adaptation to the special conditions of the Hawaiian environment. In time they became endemics, genetically distinct forms found nowhere else in the world. A few, such as the tarweeds, honeycreepers, and fruitflies, subsequently divided into many species, each attaining its own way of life, to create the adaptive radiations that are the glory of Hawaiian natural history.

The Polynesian seafarers, arriving from the Society and Marquesas Islands, broke the crucible of evolution. By introducing pigs, rats, domestic plants, and other organisms already widespread through previously occupied islands of the central Pacific, they raised the biological colonization rate thousandsfold. The alien invasion then skyrocketed when American and other settlers imported legions of additional species not just from the neighboring archipelagoes but from all over the world. Birds, mammals, and plants were deliberately introduced for their perceived value. As a result, a majority of resident land birds and nearly half of the plant species are now alien. Insects, spiders, mites, and other arthropods were unintended companions, arriving as stowaways in cargo and ballast. An average of 20 such species are detected in quarantine each year; a few slip through and succeed in establishing themselves. Among 8,790 insect and other arthropod species known to be resident in Hawaii in the late 1990s, 3,055, or 35 percent, were of alien origin. Of the grand total of 22,070 species of all kinds of organisms, plants, animals, and microbes recorded thus far on the land and in the surrounding shallow waters, 4,373 are alien. This is about half the 8,805 native species known exclusively from Hawaii. Moreover, the aliens dominate in numbers of individual organisms, especially in the disturbed environments. As a result, immigrants own the bulk of Hawaii.

Most of the invaders are relatively innocuous: only a small fraction build populations large enough to become agricultural pests or harm the natural environment. But the few that do break out are capable of enormous damage. Biologists cannot yet predict which immigrants will upon arrival become "invasives," as harmful alien species are now officially called by U.S. federal agencies. In their native habitats on distant continents, they are almost always unobtrusive, hemmed in by predators and other enemies that coevolved with them since their evolutionary birth. Now freed from these constraints in the long-sequestered and gentle environment of Hawaii and enjoying extraordinary reproductive success, they variously choke, consume, pauperize, and crowd out native species too weak to resist.

The arch-destroyers of the Hawaiian biota among the nonhuman immigrants are the African big-headed ant *(Pheidole megacephala)* and feral strains of the common pig *(Sus scrofera)*. The ant lives in borderless supercolonies of up to millions of workers and breeding queens. Spreading out as a living sheet from the point of entry, the supercolonies eat or drive out a large part of the other insects in their path. The workers are divided into two castes: slender minors that forage singly and in columns over the ground, and large-headed soldiers that use their massive head muscles and sharp mandibles to dismember enemies and prey. *Pheidole mega-cephala* is credited with exterminating most of the native insects of the Hawaiian lowlands, including those that serve as pollinators of the native flora. The impact has rippled up the food chain. By reducing the food source of some of the smaller insect-eating native birds, it has likely contributed to their disappearance. In scattered localities not occupied by African big-headed ants, supercolonies of another alien species, the Argentine ant *(Line-pithema humile),* dominate the ground in a similar way, employing mass attacks and glandular poisons to overcome opponents. Where the big-headed and Argentine ants meet, their legions fight for control of the earthen microenvironment and end up dividing the land between them. A few kinds of flies, beetles, and other insects are able to survive the combined onslaught, but in most cases they turn out to be immigrants themselves. The ants of Hawaii, like the humans of Hawaii, are aliens ruling over an impoverished dominion of fellow aliens.

The vulnerability of the Hawaiian fauna to the invasive ants conforms to a familiar principle of evolution. Almost everywhere in the world, and for tens of millions of years, ants have been the leading predators of insects and other small animals. They are also among the preeminent scavengers of dead bodies, and as turners of the soil they equal or surpass the earthworms. Prehuman Hawaii, because of its extreme isolation, never had them. In fact, no native ant is known from the central Pacific anywhere east of Tonga. As a consequence Hawaii's plant and animal communities evolved to conform to an ant-free world. They were unprepared for the shock of occupation by social predators of such high caliber. As a result, a large and still imprecisely measured part of the native Hawaiian species succumbed to their invading swarms.

The Hawaiian environment was also evolutionarily unprepared for the coming of ground-dwelling mammals. Only two mammalian species of any kind lived on the islands in prehuman times:

the native hoary bat and the Hawaiian monk seal. They have since been supplemented by forty-two alien species, all in one way or another a threat to the native Hawaiian fauna and flora. Especially destructive has been the domestic pig, carried to the islands by the early Polynesians. Some individuals escaped, or were deliberately released, and thereafter became the first large mammals to enter the forest environment. Today their feral descendants are more like the wild boars of Europe than the gentle domestic stocks from which they evolved. About 100,000 strong, they work back and forth across the woodland floor, eating the bark and roots of plants and uprooting and toppling tree ferns. The fall of the latter small trees opens the canopy and admits unaccustomed amounts of sunlight to the forest floors, altering the soil ecosystems. As they forage, the pigs sow the seeds of alien plants in their droppings, and the resulting growth crowds the native species. The pigs also dig wallows that collect pools of water. The only native insects that benefit from the standing water are damselflies, whose nymphs are aquatic. And, as always, there is a trade-off to even that small advantage. The pools are also the breeding grounds of mosquitoes, which transmit avian malaria to the genetically unprotected native birds.

People brought pigs to Hawaii deliberately, and only people can halt the destruction they cause. Teams of pig hunters, working with specially trained dogs, have cut the populations in nature reserves considerably but are nowhere near to eliminating them. In the Hawaii Volcanoes National Park of the big island, for example, about four thousand were still at large as late as 2000.

Other introduced mammals escalate the damage. Rats, mongooses, and feral house cats hunt the Hawaiian forest birds. Goats and cattle graze away the final remnants of the native vegetation in open habitats. The last individuals of some plant species survive only on inaccessible cliff faces, yet even there they are endangered by falling soil and rocks loosened by animals feeding along the cliff edges above.

Because of the relative simplicity of Hawaii's environment, it serves as a laboratory for the forces that hammer nature everywhere in the world. Among the lessons learned is that the decline of any particular species rarely has a single cause. Typically, multiple forces entrained by human activity reinforce one another and either simultaneously or in sequence force the species down. These factors are summarized by conservation biologists under the acronym HIPPO:

> *Habitat destruction.* Hawaii's forests, for example, have been three-fourths cleared, with the unavoidable decline and extinction of many species.
> *Invasive species.* Ants, pigs, and other aliens displace the native Hawaiian species.
> *Pollution.* Fresh water, marine coastal water, and the soil of the islands are contaminated, weakening and erasing more species.
> *Population.* More people means more of all the other HIPPO effects.
> *Overharvesting.* Some species, especially birds, were hunted to rarity and extinction during the early Polynesian occupation.

The prime mover of the incursive forces around the world is the second P in HIPPO—too many people consuming too much of the land and sea space and the resources they contain. To date about 205,000 species of plants, animals, and microbes have been recorded as free-living in the United States as a whole. Recent studies of the best-known, or "focal," groups, including vertebrates and the flowering plants, have revealed that the forces other than human population growth descend in order of importance in the same sequence as the HIPPO letters, from habitat removal as the most destructive and overharvesting the least. In Paleolithic times, when skilled hunters killed off large mammals and flightless birds, the sequence was roughly the reverse, OPPIH, from overharvesting to a still proportionately small amount of habitat destruction. Pollution was negligible and invasive species probably important only on small islands. With the spread of Neolithic cultures and agriculture, the sequence reversed. The newly configured HIPPO became the monster on the land, and eventually in the sea as well.

Conservation biologists, focusing on the overall problem of nature's decline, have begun to work out what appear to be countless ways that variations of the HIPPO forces join to weaken and extinguish biological diversity. Each case is a result of the unique characteristics of the threatened species and the particular corner into which human activity has pushed it. Only by focused study are the researchers able to diagnose the cause of endangerment and devise the best means to nurse the species back to health.

No species better illustrates how peculiar, even bizarre, the causes of decline can be than the Vancouver Island marmot *(Marmota vancouverensis)*. This beautiful ground squirrel was never abundant. In the late twentieth century it began a dangerous decline. By 2000 it had fallen to seventy individuals in the wild, making it Canada's most endangered species and one of the rarest animals in the entire world. Bundled in fluffy chestnut-and-white fur, with the habit of standing erect on its hind legs to survey its surroundings, it has become a charismatic species, Canada's equivalent of China's giant panda and Australia's koala bear. In the 1990s its attractive appearance and the news of its plight roused public opinion and set in motion the steps that may save it.

Field biologists searching for the cause of the decline of the Vancouver Island marmot were at first puzzled. No obvious change could be found in the immediate environment that sets the species at risk. The marmots live on the high peaks of the Vancouver Island Ranges amid rock cliffs, summer snow patches, and elfinwood fir trees scattered across subalpine meadows. Because of the remoteness of the habitat, humans rarely disturb the marmots. Nor do they hunt them. No disease appears to have recently swept the population, although it cannot be ruled out entirely as a contributing factor. The major predators of the marmot—wolves, cougars, and golden eagles—also play a role, but have been present for millennia without extirpating the species.

The problem, it turns out, is the clearcutting of forest to harvest timber *below* the subalpine habitat. With this change, an instinct that once kept the Vancouver Island marmot alive is now its undoing. Under natural conditions the small local populations composing the species frequently decline to extinction. But the empty habitats they leave behind are soon replenished, because young marmots in other populations instinctively emigrate from their homes upon reaching maturity. They move down the mountains and travel through the conifer forests that clothe the lower elevations. As they proceed they zigzag up and down the wooded slopes until they encounter another subalpine meadow. There they halt and dig burrows. The rigidity of this instinct puts the young marmots at risk in disturbed environments. When they encounter the open spaces of clearcut conifer forests, they automatically accept them as natural meadows and settle down. And there they perish, either from the more dangerous predators that prowl the lower slopes or, more slowly, from the failure of their hibernation cycle to adapt to the new temperature and snowfall regimes. Enough of the false meadows have been created by humans to bleed the more remote populations of the Vancouver Island marmot to near extinction. The concentration of emigrants in the clearings is also likely to have unbalanced the population cycles of the source populations hard by. The only way to save the species, it now appears, is to capture a few of the survivors and breed them in captivity. That rescue operation has, in fact, begun and at this writing is proving effective. At a later date the species, it is hoped, can be resettled in natural subalpine pockets surrounded by protected conifer forest.

An equally unpredictable series of mishaps earlier devastated land snail faunas of islands in the Pacific and Indian Oceans. In the early 1900s giant land snails from Africa *(Achatina fulica)* were introduced widely to serve as garden ornaments. The huge mollusks soon multiplied out of control, consuming native snails and attacking crops. In the 1950s an attempt was made to combat the *Achatina* by introducing the predatory rosy wolfsnail *(Euglandina rosea),* a native of the southeastern United States and tropical Latin America. The exercise was supposed to be a model case of biological control, in which a harmless species is introduced to whittle down a harmful one. Instead, it triggered an extinction avalanche. The rosy wolfsnails, soon to be dubbed "cannibal snails" in Hawaii, paid relatively little attention to the intended prey. Unexpectedly, they attacked and proceeded to eat their way through the native snails, which are much smaller and more vulnerable than the giant snails. To date they have extinguished more than half of the fifteen or so species of beautiful banded tree snails *(Achatinella)* native to Hawaii, and half of the closely related Hawaiian *Partulina* tree snails. They joined rats, shell collectors, and deforestation as prime agents in wiping out 50 to 75 percent of the 800 Hawaiian snail species living on the ground. They have also been complicit in the extinction of 24 of the 106 snail species endemic to the Indian Ocean island nation of Mauritius. On Moorea in French Polynesia, the rosy wolfsnails were the main cause of extinction of all of the seven free-living endemic *Partulina* snails, whose multicolored, acorn-sized shells were once used by the native people to string leis.

In a last-minute rescue operation, two biologists, James Murray and Bryan Clarke, transferred living samples of the species to several universities and zoological parks in the United States and England. Fortunately the little snails adapted well to captivity, where they can be easily housed in plastic containers and bred on lettuce. By the mid-nineties the captive populations of three species were strong enough for individuals to be reintroduced to small compounds in the rainforests of Moorea. Electric wires and moats of repellent along the boundaries protect them from the prowling rosy wolfsnails. However, one of the seven species, *Partulina turgida,* failed even in captivity. The last individual, nicknamed "Turgie" by its keepers at the London Zoo, died from a protozoan infection just ten years after the last of its kind disappeared on Moorea. Turgie's keepers constructed a little memorial on behalf of its species inscribed as follows:

1.5 MILLION YEARS B.C. TO JANUARY 1996

Among the most massive losses of recent decades has been the frog die-off. In the 1980s zoologists became aware that many amphibians around the world, principally frogs but also salamanders, were in steep decline. An early warning sign was the disappearance of Australia's unique northern gastric breeding frog *(Rheobatrachus vitellinus)*, which incubates its eggs in its stomach and gives birth to the young through its mouth. Discovered as a new species in the Eungella National Park of Queensland in January 1984, it declined and vanished by March of the following year. Simultaneously, other local Australian frog populations disappeared, following precipitous declines as short as four months. On the other side of the world, the golden toad of Costa Rica *(Bufo periglenes)* population also plummeted. Its color was spectacular: males in the breeding season looked as though they had been dipped whole in orange Day-Glo paint. That, and the toad's dramatic emergence in huge numbers during the spring breeding season, made it a zoological celebrity and important wildlife attraction for the small Central American country. In the spring of 1987 hundreds of thousands of breeding toads made their annual appearance on schedule in the only place the species occurred, the mountain forest of Monteverde. The following year a team led by David Wake of the University of California, Berkeley, could find only five individuals. No golden toad has been seen since, and the species is presumed extinct.

Meanwhile, reports poured in of the decline of amphibians elsewhere. Especially troubling was the local extinction of many species from widely scattered locations in Central and South America. Herpetologists responded with field research and conferences. In 2000 a team led by Jeff E. Houlahan of the University of Ottawa analyzed data collected by scientists over a period of decades from 936 populations in 37 countries, mostly in Europe and North America. They concluded that the amphibian numbers overall have been dropping at the rate of about 2 percent per year since at least as far back as 1960. The losses are not geographically even. Within particular regions they occur in some species and not others. For example, Canada has witnessed a 60 percent reduction in range of the leopard frog *(Rana pipiens)*, including its complete disappearance from British Columbia. In California, all of the frog species of Yosemite National Park have declined. The mountain yellow-legged frog *(Rana muscosa)* has vanished from the western slopes of the Sierra Nevada but continues strong on the eastern slopes. Frogs and salamanders of the southeastern United States, one of the world centers of amphibian biodiversity, have so far held up relatively well.

As researchers focused on what they named the Declining Amphibian Phenomenon, they came to agree that the prime mover is the destruction of habitat, the H in the aforementioned HIPPO. But a medley of other malign forces are at play, sometimes stemming directly from habitat loss, sometimes independent of it. These forces press down in permutations that differ from one locality to the next and according to circumstance. In the Sierra Nevada, atmospheric pollution from the coast evidently plays a role. To the north, in the Cascade Mountains of Oregon, a leading culprit is the increase of ultraviolet-B radiation, the cell-damaging component of sunlight. That boost is due, in turn, to the thinning of the Earth's ozone layer, another human-caused alteration of the environment, and one most severe in high latitudes. Elsewhere in the western United States, trout and bullfrogs introduced into streams feed voraciously on smaller frog species and account for some of their extirpations. In Minnesota many leopard and cricket frogs flop around with extra limbs or missing hind legs. The growth abnormalities are believed to be caused by chemical pollutants; among the prime suspects is methoprene, sprayed in the water to prevent growth of mosquito larvae. In Central America the principal killer is almost certainly a microscopic chytrid fungus, which massively infects the soft skins of the frogs. Because frogs must breathe through their skins, they suffocate. The fungus has been spread, at least in part, in water carried in aquaria from one country to another.

Through their suffering the frogs give us a sharp, clear warning of HIPPO's lethal erosion of the biosphere. Frogs are nature's canary in the mine. The adults of most species are unusually sensitive to small changes in the environment because they live in the water, permanently or part-time, or in moist hideaways in forests. Their larvae—tadpoles—are aquatic bottom feeders. In both adult and larval stages the typical amphibian skin functions as a moist and porous apparatus for the exchange of air, making it an absorbent pad ideal for collecting poisons and parasites. We ourselves could not have devised a better early-warning device for general environmental deterioration than a frog.

The amphibians illustrate another principle pertaining to the health of biodiversity: species stressed by HIPPO forces become more vulnerable to natural death. The most insidious of such lethal agents is inbreeding depression. The smaller the population, the higher its level of inbreeding—that is, the more frequently close relatives such as siblings and first cousins meet and mate. The higher the level of inbreeding, the larger the percentage of offspring in the population with double doses of defective genes that cause sterility and early death. Inbreeding depression has been measured in laboratory analyses of fruitflies and mice. It has been documented in wild populations of greater prairie chickens (*Tympanuchus cupido*) in Illinois, as well as Glanville fritillary butterflies (*Melitaea cinxia*) in Finland. It is undoubtedly a common phenomenon in rare plant and animal species everywhere. As a rule, inbreeding starts to lower population growth when the number of breeding adults falls below five hundred. It becomes severe as the number dips below fifty and can easily deliver the coup de grâce to a species when the number reaches ten.

Inbreeding depression is not, however, an inevitable consequence of small population size. If the species manages to pass through a bottleneck of very low population size and still survive, the depression may in the course of the passage "clean out" the defective genes. Such a genetic purge evidently occurred in the cheetah. This graceful African cat, the world's fastest animal on the ground, is endangered largely due to low infant survival. In one study of the Serengeti population, 95 percent of the cubs failed to reach independence a year after birth. But they did not perish from genetic defects, as might be immediately suspected. The principal causes instead were predation by lions and spotted hyenas, along with abandonment by the mothers during periods of food scarcity.

Rarity hurts in other ways. Below fifty individuals the relative degree of random fluctuation in population size increases, and the up-and-down demographic bounce can easily reach what mathematicians call the "absorbing barrier"—zero, point of no return.

A very small or very local population is also set up for near-instant demise from any storm, flood, wildfire, drought, or other natural disaster that comes along. Such was the story of the recent brush with extinction of one of America's most beautiful butterflies, Schaus' swallowtail *(Heraclides aristodemus ponceanus)*. Once common on the mainland of South Florida and the northern Florida Keys, this large lepidopteran with the chestnut-and-amber wings became increasingly restricted by the clearing of its forest habitat. It was diminished further by insecticide sprays used to control mosquitoes. By 1992 it could be found only on four keys, all within Biscayne National Park and the northern tip of Key Largo. On August 24, 1992, Hurricane Andrew, one of the most destructive storms in modern American history, swept through the area, devastating all five of the remnant habitats and driving Schaus' swallowtail to the brink of extinction. A small captive breeding population is now maintained by the entomologist Thomas Emmel at the University of Florida in Gainesville, a slender buffer against total extinction of the race.

If the extinction of a species is a sniper shot, then the destruction of a habitat containing multiple unique species is a war against nature. The clearcutting of a remnant patch of mountain rainforest can eliminate scores of species in one stroke. Such a catastrophe occurred, for example, when farmers cleared Ecuador's Centinela Ridge during 1978–86, extinguishing as many as seventy kinds of plants found nowhere else. Aquatic equivalents of the Centinela massacre are especially common. The freshwater mussel fauna of the United States, comprising 305 native species, is one of the richest in the world, but widespread damming and pollution of the nation's rivers and tributary streams has reduced it by more than 10 percent. Half of the survivors are imperiled to some degree, and of these half again are classified as critically endangered, one short step from oblivion.

Of all forms of ongoing habitat destruction, the most consequential is the clearing of forests. The maximum extent of the world's forests was reached six thousand to eight thousand years ago, at the dawn of agriculture and following the retreat of the continental glaciers. Today, due to the universal spread of agriculture, only about half of the original forest cover remains, and that is being cut at an accelerating rate. Over 60 percent of temperate hardwood and mixed forest has been lost, as well as 30 percent of conifer forest, 45 percent of tropical rainforest, and 70 percent of tropical dry forest. As recently as 1950 Earth's old-growth woodland occupied 50 million square kilometers, or nearly 40 percent of the ice-free surface of the land. Today its cover is only 34 million square kilometers and is shrinking fast. Half of that surviving has already been degraded, much of it severely.

The loss of forest during the past half-century is one of the most profound and rapid environmental changes in the history of the planet. Its impact on biodiversity is automatic and severe. To reduce the area of a habitat is to lower the number of species that can live sustainably within it. More precisely, as the area shrinks, the sustainable number of species falls by the sixth to third root of the area. A common intermediate value found in nature is the fourth root. At the fourth root, the reduction of a habitat to one-tenth its original area eventually causes the fauna and flora to decline by about one-half. The classic example of the principle is the drop in the number of species of reptiles and amphibians found on various West Indian islands in descending order of size, starting for example with Cuba (44,164 square miles, about 100 species) and proceeding downward through Puerto Rico (3,435 square miles, 40 species), Montserrat (33 square miles, 25 species), and finally to tiny Saba (5 square miles, 10 species) and Redonda (1 square mile, 5 species).

The same rule applies to the national parks of the western United States and Canada. While not traditional islands surrounded by the sea, like the West Indies, they nevertheless are "habitat islands" in a sea of ranches, farms, and cutover timberland. During the hundred years of their existence, the number of mammal species has declined in accordance with the mathematical deductions of island biogeography. Also as predicted by theory, the smaller the park, the steeper the drop. The largest reserves, such as the joint Glacier National and Waterton-Glacier International Peace Park in Montana and Alberta, have yet to lose a single species.

A frightening aspect of the area-species principle is that while removal of 90 percent of the habitat area allows about half of the species to hang on, removal of the final 10 percent can wipe out the remaining half in one stroke. In fact, the number of natural habitats reduced to fragments this size or smaller is increasing rapidly all around the world.

The headquarters of global biodiversity are the tropical rainforests. Although they cover only about 6 percent of the land surface, their terrestrial and aquatic habitats contain more than half the known species of organisms. They are also the leading abattoir of extinction, shattered into fragments that are then being severely adulterated or erased one by one. Of all ecosystems, they are rivaled in rate of decline only by the temperate rainforests and tropical dry forests. Since the 1980s, according to estimates of the Food and Agriculture Organization of the United Nations (FAO), the worldwide rate of clearcutting (reducing local forests to 10 percent of original cover or less) has been close to 1 percent per year. Where all tropical rainforests together occupy an area approximately equal to the lower forty-eight United States, they are being removed at the rate of half the state of Florida annually. There follows, for example, the rates of tropical deforestation for individual countries of South America during 1980–90, as calculated by the FAO in square kilometers:

	TROPICAL FOREST AREA REMAINING IN 1990 (KM²)	ANNUAL DEFORESTATION RATE, 1980–90 (KM²/YEAR)	ANNUAL DEFORESTATION RATE, 1980–90 (PERCENTAGE)
Bolivia	459,000	5,320	1.16
Brazil	4,093,000	36,710	0.90
Colombia	541,000	3,670	0.68
Ecuador	120,000	2,380	1.98
Peru	674,000	2,710	0.40
Venezuela	457,000	5,990	1.31

A few experts, including the British ecologist Norman Myers, believe the damage to tropical forests has been underestimated by the FAO, with the true figure being closer to 2 percent per year, or the equivalent of all of Florida. On the other hand, recent satellite data support a lower rate that, for South America at least, indicates that the FAO estimates are too high by a factor of as much as two. Bolivia, according to these studies, was being deforested in 1986–92 at the annual rate of 0.52 percent, and Brazil, from one year to the next during 1988 and 1998, variously between 0.30 and 0.81 percent per annum.

Of twenty-five "hotspots" on the land—places with the most species at risk of extinction—fifteen are covered primarily by tropical rainforests. These threatened ecosystems include the moist tropical woodlands of Brazil's Atlantic coast, southern Mexico with Central America, the tropical Andes, the Greater Antilles, West Africa, Madagascar, the Western Ghats of India, Indo-Burma, Indonesia, the Philippines, and New Caledonia. Together with the remaining terrestrial hotspots, covered mostly by savanna and coastal sagebrush, they take up only 1.4 percent of the world's land surface. Yet, astonishingly, they are the exclusive homes of 44 percent of the world's plant species and more than a third of all species of birds, mammals, reptiles, and amphibians. Almost all are also under heavy assault. The rainforests of the West Indies, Brazil's Atlantic coast, Madagascar, and the Philippines, for example, retain less than 10 percent of their original cover.

Large numbers of species have already been lost forever from the forest hotspots. Many more are endangered. In a nightmare scenario, battalions of loggers armed with bulldozers and chainsaws could wipe these habitats off the face of Earth in a few months—and with them a large part of the world's biodiversity. On the flip side, it is heartening in compensating degree to realize that by protecting this tiny fraction of the planet's land area, millions of species can be saved for posterity.

Also in balance are the remaining wildernesses, or frontier forests, as they are often called: the vast rainforests of Amazonia, central Africa (including especially the Congo Basin), and New Guinea, and the needleleaf forests of Canada and the Russian Federation. Once the legendary strongholds of Malaysia, Sumatra, and Borneo were in this group, but they have been damaged so extensively in the past several decades as to compose forests of less than wilderness quality.

The frontier forests have survived into the twenty-first century frayed, transected, and pockmarked, but still relatively intact. Earth's greatest single reserve among them is the Amazonian forest, which is larger than the Congo and New Guinea forests combined. To traverse this wonder by air, watching its green carpet slide beneath, bounded only by the horizon, the surface sparkling with sunflashes from rivers and oxbow lakes, is to renew hope for the salvation of the living world. To visit it on foot exactly as it was experienced by Humboldt, Darwin, and Bates, and Amerindian hunters millennia before them, where in ten square kilometers may be found more kinds of plants and animals than in all of Europe, is an opportunity I hope will be kept open for all generations to come.

Vast it is, but the Amazon rainforest is not safe. The countries that own the wilderness are inclined to treat it as a timber resource and promised land for the rural poor, while corporate strategists foresee immense profits from tropical crops grown in the cutover soil. Given that the trees are anchored only by shallow root systems and can be easily bulldozed over, then sawn into lumber, chipped, or burned, the Amazonian wilderness could be easily destroyed within decades. Approximately 14 percent of its cover has been converted already. Brazil, which owns two-thirds of the Amazon and other South American tropical rainforest, has set aside only 3 to 5 percent of the cover in fully protected reserves. An ultimate goal recently established by the Brazilian government is 10 percent.

Ten percent is not enough to save the Amazon as we know it. It would not protect the vast assemblage of plants and animals that make Brazil the biodiversity capital of the world. In theory, about half the species could survive indefinitely on this amount of land. But tropical forests, in spite of their fabled exuberance, are more vulnerable and less resilient than most other ecosystems. Their weakness lies in the nutrient-poor soil, which surrenders easily to the erosive power of heavy rain. The hardwood and needleleaf woodlands of the north temperate zone are grounded in deep layers of humus, where seeds lie dormant for years. When the trees are clearcut, and if the soil stays mostly intact, the original woody vegetation quickly grows back. Even soil that has been plowed for generations can often support a nearly full and early regeneration. Not so in most of the Amazon. Imagine walking with shovel in hand into a typical Amazon terra firme forest, away from and a bit above the floodplain. In deep shade from the canopy high above, working around tangles of vines, understory palms, and flank buttresses of great trees, you find an open space and dig into its soil. In one cut you are through the litter and humus; the organic material has mostly petered out at an inch or two beneath the surface. Here and there are patches completely bare of litter and humus, as though swept clean by a broom. Now look at the trees and their dense canopy. This is an ecosystem where the biomass is locked up in aboveground organisms. Dead vegetation falling to the ground has no time to pile up. It is quickly broken into shapeless frass by legions of arthropods, annelid worms, fungi, and bacteria. The last-stage nutrients released by these ministrations are quickly absorbed by the shallow rootlet masses of the trees and understory shrubs. When small sections of forest are opened by natural treefall or small-scale swidden agriculture, the thin humus holds in place and the gap is closed by regeneration from the nearby surrounding forest. But when large swaths of forest are clearcut and burned, the usual practice, most of the humus cover is too far away to be reseeded quickly, and the pounding rains soon wash it away.

The tearing down of a tropical rainforest follows a typical sequence. First, a road is cut deep into the interior for logging and settlement. Trails and forest camps are built along the way, and hunters fan out to furnish game—"bushmeat"—for the work crews. The land, depleted of the best timber and in declining condition, is next sold in lots to ranchers and small farmers. They build small roads laterally off the main route, creating the fishbone pattern seen, for example, along the highways west through Rondônia and north from Manaus to Boa Vista near the Guyana border. The settlers topple most of the rest of the trees, use some for lumber, and let the rest dry out. A year later they burn the debris. The ash that settles on the barren soil is enough for several years of good harvest. As the ground nutrients wash away, the settlers either make the best of a worsening situation or else move on to occupy new land closer to the frontier. Some are lucky enough to find tracts of deeper, more persistent and fertile soil. Repeating the ruinous cycle endlessly, the human wave is rolling up the Amazon forest like a gigantic carpet.

Within the landscape fish bones, the destruction is not immediately total. As the settlers press outward they leave bits and pieces of forest here and there, little refuges along riverbanks, on steep slopes, and in watery sloughs. But these fragments are far from secure refuges for the bulk of Amazonian life. The larger mammals and palatable birds are quickly hunted out, creating what conservation biologists call the silent forest syndrome. In one field trip near Manaus I spent my time in such a 2.5-acre fragment, content with the still-impressive variety of ants I could find, but knowing I had no chance of encountering a jaguar, a troop of howler monkeys, or a peccary herd. Such fragments, even if left untouched, do not carry on as microcosms of the pristine larger forest. Cut out from the deeper reaches as though by a giant cookie cutter, with no protective marginal vegetation left in place, they suffer the edge effect, a kind of forest sickness. Winds blow in from the side, drying out the interior to distances of a hundred meters or more. In the peripheral zone, ground plants adapted to deep forest decline and die out. The towering trees above them are weakened. During the violent storms common in the region, gusts of wind snap off branches or even full crowns. Some of the trees fall entire, knocking down other trees in their paths, and decapitating others by the hawserlike vines that grow from one crown to the next. Treefalls occur in undisturbed forest too; you can occasionally hear them on a quiet day from a mile away. They create tree gaps that are part of the normal cycle of growth, small enough to receive seeds from the surrounding forest. The saplings and herbaceous ground vegetation that spring up in the gaps add to the diversity of primary forest. In the edge zone of a forest isolated by extensive clearcutting, however, this process is dangerously accelerated. It opens wider spaces, exposing the tree bases to more sunlight, killing the shade-loving epiphytes, drying the soil and litter, and admitting a swarm of weedy plants and animals. As a result, a fragment of forests even as large as 1,000 hectares (2,500 acres) can in time consist entirely of altered habitat.

In the damaged portions of the still-standing forest the stage is set for the arrival of an even more potent menace. Fires, ignited by lightning strikes or set by farmers to remove brush, sweep through the desiccated outer zones. Building momentum, they work on into the deep interior.

Roads and small settlements that penetrate a pristine forest fragment it and commence its conversion before the clearcutting that follows. This early damage is difficult to detect from the air and may be invisible to remote sensing by satellites. It is best assessed on the ground. In 2000 over 40 percent of the Amazon cover was estimated to have been disturbed to some degree.

In time the transformation reaches a critical point and becomes self-sustaining. As the early damage spreads, new intrusions add to the impact and reinforce one another, a process environmental scientists call synergism. During droughts in the El Niño cycle, the forests burn more fiercely than in other years. In 1998 smoke plumes from myriads of forest fires grew so thick that airports at the Amazonas capital of Manaus and elsewhere downwind were closed for short periods of time. When such smoke becomes thick enough, it kills seedlings. It can even choke off rainfall: the fine particulate matter of smoke rising into the air forms so many condensation nuclei that atmospheric water remains as mist and cannot aggregate into drops large enough to fall to the trees and soil.

An equally harmful synergism is the reduction of atmospheric water over the Amazon that comes from the trees themselves. The potential exists for a climatic downward spiral: cut down trees, reduce rainfall, lose still more trees. About half of the rain falling in the Amazon Basin arises from the forest itself, as opposed to clouds that blow in from rivers and the Atlantic. The water is carried in conducting vessels of the plants and evaporated from their leaves and branches. To the extent that the Amazon is diminished by cutting and burning, annual rainfall also declines, and the wilderness remnants are stressed still more. Mathematical models of the process suggest that a tipping point exists that in the future could cause the forest ecosystem to collapse, yielding much of its area to dry scrubland.

The same principles apply to moist tropical forests elsewhere. Those of Indonesia may be closely approaching the critical damage levels predicted by theory. Eighty percent of the forest cover has been committed to logging and replacement by oil palm and other plantations, and rapid clearing is under way. The result, intensified by an already strong drought, has been some of the worst forest fires in the history of tropical Asia. In 1997–98 alone about 10 million hectares (25 million acres) went up in smoke. Even parts of the interior previously too wet to burn were lost. Most of the forests of this region, including 15 million hectares (37 million acres) on the island of Borneo, have an intrinsic ecological weakness that exacerbates the damage. They are composed primarily of dipterocarp trees, which flower during El Niño years and disperse their seeds during a six-week period. The heaps of seeds that fall to the ground are a bounty for deer, tapirs, porcupines, orangutans, birds, insects, and other animals of diverse kinds. After gorging themselves, these creatures still leave enough seeds to sprout the next generation of dipterocarp seedlings. In Indonesian Borneo many of the dipterocarp species have failed to reproduce at all since 1991, even in completely protected reserves.

In short, the damage inflicted on the great Asian forests by human activity has turned El Niños from creators into destroyers. These events are part of a natural climatic cycle called the El Niño Southern Oscillation—ENSO for short—during which tropical surface waters of the ocean are alternately warmed (the El Niño phase) and cooled (the La Niña phase). The effect on climate varies in detail from region to region, but on a more global scale it first raises and then lowers temperature and rainfall, while increasing the frequency and intensity of storms. The cost of ENSOs to natural environments already weakened by human activity can be devastating.

In recent years ENSOs have grown in frequency and amplitude. It is tempting to link the trend to global warming, as some experts have done. That conclusion is not firm, however. The mathematical models of climate change employed to the present year (2001) have not focused on small enough sections of the ocean surface either to allow or to discount it. Still, even if there is little or no enhancement of the ENSO events, climate models for the twenty-first century project with 66 to 90 percent likelihood greater ENSO-related floods and droughts around the planet.

Meanwhile, there can no longer be any reasonable doubt of global warming itself and its generally malign consequences for the environment and human economy. According to estimates based on tree rings, fossil air samples trapped in glacial ice, and other proxies, the mean surface temperatures of Earth varied by less than 2°F during most of the ten thousand years following the end of the Ice Age. Then, from 1500 to 1900, it rose approximately 0.9°F, and from 1900 to the present time it has increased another 0.9°F. The most authoritative studies of this trend are those conducted by the Intergovernmental Panel on Climate Change (IPCC), the more than one thousand experts worldwide who specialize in different aspects of the phenomenon. In 2000 they affirmed that global warming, as suspected earlier, is indeed caused principally by the heat-absorbing greenhouse gases carbon dioxide, methane, and nitrous oxide. For the past 400,000 years, the span for which reliable estimates can be made from chemical measurements of trapped glacial air bubbles, carbon dioxide concentrations have fluctuated in lockstep with surface temperature. The concentrations of carbon dioxide are now at the highest level recorded for the 400,000-year span and show no sign of leveling off. The same is true of methane and nitrous oxide. It is further reasonably certain that the thickening of the greenhouse gases is due in good measure to industrial activity and the cutting and burning of forests.

In 1995 the IPCC scientists, working with the most advanced computer programs of the time, projected that the acceleration of global mean surface temperature will continue, resulting in a further increase of between 1.8°F and 6.3°F by the year 2100. Their conclusions, and recommendations for mitigation, resulted in he 1997 Kyoto Protocol, an international treaty aimed at cutting greenhouse gas emissions by 5.2 percent within a decade. The latest models, released in early 2001, predict that if nothing is done, Earth's mean global surface temperature will rise by at least 2.5°F and as high as 10.4°F within this century. (The range reflects uncertainties in future population growth, consumption, and energy conservation.) Even if fully implemented, the Kyoto Protocol will shave that increment by as little as 0.1°F. Could these predictions be wrong? We prayerfully hope so, but with each passing year they are more solidly grounded, to the extent that it would be criminally negligent to ignore them. In ecology, as in medicine, a wrong diagnosis can cause far more harm if it is negative than if it is positive.

More frequent heat waves, violent storms, forest fires, droughts, and flooding are the spawn of the historically unprecedented pace of climate change. Polar ice caps are destined to weaken: in the summer of 2000 an icebreaker made it through thin and disappearing ice all the way to a mile-wide pool of open water at the North Pole. Sea levels, if the trend continues, will rise by four to thirty-six inches. Shallow coastlines around the world will shrink. In the Pacific and Indian Oceans, many atolls, including the small island nations of Kiribati, Tuvalu, and the Maldives, will partially disappear. Real-estate investment in New Orleans and the Florida Keys, not to mention the Bahamas and New York City, will seem an increasingly poor long-term risk.

As the thermoclines of global climate ease poleward, plant and animal species will be hard-pressed to keep up. Nine thousand years ago, as the continental ice sheet retreated across North America at a velocity of 120 miles a century, two species of cold-loving spruce successfully followed close behind. Today they help fill the great needleleaf forests of glacier-free Canada and Alaska. But the ranges of most other tree species poked along at only five to twenty-five miles per century. With an even faster northward velocity of climatic zones projected for the twenty-first century, the future of slower-paced native floras and faunas in the temperate zones is problematic. Many native species are already trapped in natural reserves that have become islands in a sea of cropland and suburban sprawl. Others, as in Florida, face a different risk: they are limited by their genetic adaptation to coastal strips likely to be submerged by the rising sea.

Some of the North American species threatened by climate change can be transplanted northward or inland. But elsewhere in the world there are ecosystems with no place at all to go. Among the most extensive are the tundras and seas of the high latitudes. With even a modest amount of global warming they will be pushed toward the poles and then oblivion. Thousands of species, from lichens and mosses to penguins, polar bears, and reindeer, could be lost. The same fate awaits arctic-alpine biotas of the high mountain ranges and the upper montane tropical rainforests in other parts of the world.

Back-to-the-wall entrapment also threatens the faunas and floras of the Gondwanan lands. These biologically unique worlds form a broken ring composed of the southern strips of Earth's ice-free land. They include the cool temperate reaches of southern South America, southernmost Africa, Madagascar, Antarctica, the subantarctic shelf islands, the Indian subcontinent with Sri Lanka, Australia, and the archipelagoes of New Zealand and New Caledonia. The original Gondwana, one of the two supercontinents of the ancient world (the other one, Laurasia, lay to the north), split into the present fragments during late Cretaceous times, toward the end of the Age of Dinosaurs some 100 million years ago. Once covering a large part of Earth's surface, old Gondwana has been the site of some of the earliest events of terrestrial evolution. The paleosols of South Africa have yielded what appear to be the chemical signatures of land-dwelling bacteria two billion years before the present. The evidence, if confirmed, will triple the known age of terrestrial life. Gondwana was also home to the first known terrestrial vascular plants, which originated there during the Silurian period about 425 million years ago.

The living Gondwanan life forms, some with histories dating back to the time of the supercontinent, are treasures worth saving. Unfortunately, as the climate warms and subtropical and tropical temperature regimes press southward, some cool-temperate floras and faunas have nowhere to go except into the Indian Ocean. The situation is especially threatening to southern Africa below the Cunene and Zambesi Rivers. There thirty thousand species of flowering plants live, of which over 60 percent are found nowhere else. Within this domain, even the arid habitats rank among the richest on Earth. They contain, for example, an impressive 46 percent or more of the succulent plant species of the world, forming the natural garden of the wonderfully named Succulent Karroo.

Adding to worldwide stress on the natural world from habitat destruction and climate change is the rising tide of alien species, the Hawaiian problem writ large. Most settle in urban and agricultural environments, snuggling close to the human populations that transported them. By chance, however, a small percentage are preadapted to fill whatever niches that are open or can be forced open in the surviving natural ecosystems. A very few are able to penetrate the heart of the natural environment itself, with sometimes devastating results.

According to a study by the Office of Technology Assessment of the U.S. Congress, at least 4,500 alien species of plants, animals, and microbes had been established in the United States by 1993, adding to the more than 200,000 known native species. This figure is probably a considerable underestimate. The true number, if species with even a tenuous foothold are included, could, according to a second study made in 2000, top 50,000. Some of the immigrants, such as the crop and livestock species that yield almost all of America's agricultural produce, are here with our blessing. Others, including a legion of agricultural and household pests, impose total costs approaching $137 billion each year.

Some of the immigrants gone wrong were brought to the New World with the best of intentions. In 1890–91, about one hundred European starlings were released in New York City by a collector whose goal was to establish in America all of the birds mentioned by William Shakespeare. Today they are a plague across America. Most other invasives arrived unnoticed as stowaways.

Given that the flood of immigration will continue, what will be the long-term effect? Might the resulting enrichment of the fauna and flora yield more benefit than harm to humanity? The answer from experience is almost entirely negative. Only if the influx were carefully controlled to admit a tiny number of safe and useful species, using vastly more biological knowledge than we now possess, might we tip the balance the other way. The reason has been clearly documented in case histories from around the world. In their own native land the immigrant species are held in check by natural enemies and other population controls. Released from these restraints in the new, host environment, a few explode in number and spread. While some may be beneficial in certain ways, they are almost always more than equivalently harmful in other ways. With admirable irony, ecologists have expressed this imbalance in the titles of several recent books: *Alien Invasion, America's Least Wanted, Biological Pollution, Life Out of Bounds,* and *Strangers in Paradise.* Consider the following examples from the American environment, with harm weighed against benefit.

• Chestnut fungus *(Cryphonectria parasitica).* After its accidental introduction on logs of Asian chestnut into New York City in 1904, this species blanketed over ninety million hectares of forest within fifty years. *Harm:* The fungus virtually eliminated the American chestnut, the dominant tree of the eastern American forests. It thereby altered and diminished the entire timberland environment. Almost unnoticed except by entomologists, seven moth species that fed exclusively on the chestnut became extinct. *Benefit:* None yet discovered.

• Red imported fire ant *(Solenopsis invicta).* This notorious little insect was introduced from the Brazilian-Argentine border region into the port of Mobile, Alabama, in the 1930s, probably as a stowaway in ship cargo. It then spread throughout the South, from the Carolinas to Texas. In the 1990s it established an enclave in Southern California. *Harm:* The red imported fire ant, whose sting feels like the touch of a hot needle, is a major pest of agriculture and households and a threat to wildlife. *Benefit:* It preys on and reduces other insect pests in sugarcane fields. I am inclined to think that, if they could, the farmers would choose the other insect pests.

• The Asian subterranean termite *(Coptotermes formosanus),* "the termite that ate New Orleans." *Harm:* Rapidly multiplying, voracious, hard to exterminate, this insect is responsible for hundreds of millions of dollars of damage each year from Florida to Louisiana. *Benefit:* There must be something, but so far no one has thought of it.

• Zebra mussel *(Dreissena polymorpha).* This small, banded bivalve was introduced into the Great Lakes in the 1980s from either the Black Sea or the Caspian Sea. It then spread rapidly downstream through the Mississippi Valley, eventually reaching the Gulf Coast. Most recently it puddle-jumped into New York and New England. Extraordinarily prolific, the zebra mussel forms continuous beds of shells on the floor of freshwater lakes and streams. *Harm:* To start, it clogs the intake pipes of electrical utilities, shutting them down. The cost of this and other forms of damage will, according to the U.S. Fisheries and Wildlife Service, reach a cumulative $5 billion by 2002. The zebra mussel also has eliminated some native mollusk species locally by growing over and crowding them out. Further, by filtering and clearing water of particulate matter, it reduces phytoplankton and the food supply of other filter-feeders and their predators, thereby altering entire aquatic ecosystems. *Benefit:*

Where water has been cleared by the mussels, for example in Lake Erie, aquatic vegetation grows more luxuriantly. As a result of this change, some native mollusk and fish species have increased in numbers. Putting aside the economic damage, the ultimate environmental effect of the zebra mussel invasion is hard to judge, but its aggressiveness and superabundance make it at best a high-risk environmental player. Still, why waive aside economic damage? Five billion dollars is not chump change.

• Purple loosestrife *(Lythrum salicaria)* was introduced as an ornamental from Europe to the gardens and wetlands of the U.S. Atlantic seaboard. Forming dense stands in moist soil, this perennial plant has spread aggressively across the entire northern United States, coast to coast, and into southeastern Canada. *Harm:* Known by conservationists as the "purple plague," *Lythrum salicaria* crowds out cattails and other elements of the native wetland vegetation. *Benefit:* Fields of the lovely stranger now cover much of America's semiwild landscape, their profuse tall spikes creating lovely impressionist images throughout the summer. An added touch of color might be welcome, but not, biologists and conservationists add, at the cost of the native wetland flora.

• Tamarisks (*Tamarix,* several species). Introduced from Eurasia, these small trees form groves that have become a standard feature of streamside environments of the American deserts. *Harm:* They suck up underground water with high efficiency, outcompeting native vegetation and diminishing the variety and abundance of wildlife. *Benefit:* Tamarisks are pleasant shade trees, to be savored by those unaware or uncaring that they are in a biologically impoverished environment.

• Kudzu *(Pueraria lobata).* This prodigiously adaptable leguminous vine, able to grow two inches an hour, was brought into the United States in 1876 to decorate the Japanese Pavilion at the Philadelphia Centennial exposition. *Harm:* Kudzu, justifiably called the "plant that ate the South," is an originally friendly alien run amok. It spreads swiftly, not only across naked red-clay soil but through a wide variety of other habitats from open forests to urban yards, blanketing trees, utility poles, highway signs, and small buildings. It smothers gardens and small agricultural fields. The annual cost of its exuberance has been estimated to exceed $50 million. *Benefit:* Since the early 1900s kudzu has been used as a shade plant and livestock forage crop. In the 1930s, when much of southern agricultural land had eroded into barren fields and gullies, the species performed superb service by holding and restoring soil. The United States Soil Conservation Service and a citizens' Kudzu Club promoted its propagation. Today it is viewed as a mixed blessing that we can easily live without.

• Miconia *(Miconia calvescens).* This attractive tree, a native of tropical America, was introduced as an ornamental into French Polynesia. *Harm:* Now known on Tahiti as the "green cancer," miconia escaped from cultivation to form dense stands of trees, up to fifty feet in height, that crowd out almost all other forms of plant life. This one plant now dominates about two-thirds of the island. Miconia has also become the greatest single threat among all invasive species to Hawaii's tropical forests, where it has thus far been kept under control only by intensive weeding. *Benefit:* Nice to look at, if you can keep it confined to gardens. On second thought, it is better not to plant miconia anywhere it can grow outdoors.

• Balsam woolly adelgid *(Adelges piceae).* A very small insect with a huge environmental wallop. *Harm:* This European aphid, the animal equivalent of the chestnut blight fungus, has wiped out virtually all of the adult fir trees in the Great Smoky National Park and thereby eliminated three-quarters of the spruce-fir forest cover of the southern United States. *Benefit:* None yet discovered, although suggestions will be welcomed by foresters and the National Park Service.

• Brown tree snake *(Boiga irregularis).* Introduced onto the island of Guam from New Guinea or the Solomon Islands shortly after the Second World War, this poisonous reptile is arguably the most frightening of all invasive species. *Harm:* A voracious predator of birds, growing to ten feet in length, and at one time reaching densities of up to twelve thousand per square mile, *Boiga irregularis* drove into extinction all of the ten forest species of Guam, including a rail, a kingfisher, and a fly-catcher found nowhere else. Emerging from the forest, the big snakes also invade farms and houses, where they empty chicken coops and attack family pets. *Boiga irregularis* has authorities on high alert in Hawaii, where several individuals have been inter-cepted over the last several years at the Honolulu airport. If the brown tree snake succeeds in colonizing the islands and per-forms as on Guam, it could wipe out a large section of both the introduced and native bird faunas. *Benefit:* With poisonous rep-tiles generally unpopular, and the market for snake meat still quirky and unreliable, the brown tree snake is not likely to be welcomed as an immigrant anywhere anytime soon.

Imagine the natural world a hundred years from now if current environmental trends continue:

In 2100 there are still Joseph Turner landscapes of inanimate beauty in many parts of the world. People still enjoy snow-covered mountain peaks, wave-dashed headlands, and whitewater rapids tumbling into deep, still pools. But what of living nature? The huge human population, having at last leveled off at nine to ten billion, occupies the entire habitable part of the planet, which has been turned into a tight mosaic of cropland, tree farms, roads, and habitations. Thanks to massive desalinization put in place by 2100, new methods of fresh-water transport, and irrigation, drier regions have turned from brown and yellow to green. Global per-hectare food production is well above 2000 levels. More of the 50,000 species of potentially palatable kinds of plants are in agri-cultural use, while genetic engineering has been applied to tweak older crop species toward their productive limits.

A global technoscientific civilization has risen from the cauldron of ethnic and class conflict, which nevertheless still simmers beneath it. People in 2100 are better fed and educated than in 2000, but the great majority are in the developing world and remain poor even by the standards of industrialized countries a century before. On a planet destined to remain overpopulated well into the twenty-second century, elite rich countries remain in conflict with resentful poor countries. War is rare and terrorism has diminished, but it is a tense world, still roiled by the anguishing contradictions of human nature.

Humanity is aging fast in 2100. The means have been provided to eliminate most diseases, including those caused by genetic defects. Medical delivery has improved dramatically almost everywhere. The big news is the lengthening lifespan, bought by staggering increases in the cost of medical care. Postcentenarians are commonplace. The causes of aging are known, and birthrates have plummeted in compensating degree, especially in the richest countries, where young people are increasingly obtained through recruitment from poorer countries. The genetic homogenization of the world population by intermarriage, already well advanced by 2000, has accelerated. There is more genetic variation within local populations but less between the populations than was the case back in 2000. Biological races grow fainter with the passage of each generation.

None of these changes has altered human nature in the least. No matter how sophisticated our science and technology, advanced our culture, or powerful our robotic auxiliaries, *Homo sapiens* remains in 2100 a relatively unchanged biological species. Therein lies our strength, and our weakness. It is the nature of all biological species to multiply and expand heedlessly until the environment bites back. The bite consists of feedback loops—disease, famine, war, and competition for scarce resources—which intensify until pressure on the environment is eased. Add to them the one feedback loop uniquely available to *Homo sapiens* that can damp all the rest: conscious restraint. For the trends of 2000 to have continued, as I imagine in this scenario, means that restraint has failed.

In 2100 the natural world is suffering terribly. The frontier forests are largely gone—no more Amazon or Congo or New Guinea wildernesses—and with them most of the biodiversity hotspots. Coral reefs, rivers, and other aquatic habitats have deteriorated badly. Gone with these richest of ecosystems are half or more of Earth's plant and animal species. Only a few fragments of wild habitats persist as relics here and there, guarded by governments and private owners rich and wise enough to have held them fast as the human tidal wave washed over the planet.

Like human genetic diversity, the fragmentary biodiversity that survived to 2100 has also become much more geographically simplified. The cosmopolitan flow of alien organisms has flooded each fauna and flora with immigrants from multiple other faunas and floras. To travel around the world along any chosen latitude is to encounter mostly the same small set of introduced birds, mammals, insects, and microbes. These favored aliens compose the small army of companions that travel best in our globalized commercial transport and thrive in the simplified habitats we have created. An aging and wiser human population understands very well—too late now—that Earth is a much poorer place than it was back in 2000, and will stay that way forever.

Such is likely to be the world of 2100—if present trends continue. The most memorable heritage of the twenty-first century will be the Age of Loneliness that lies before humanity. The testament we will have left in launching it might read as follows:

We bequeath to you the synthetic jungles of Hawaii and a scrubland where once thrived the prodigious Amazon forest, along with some remnants of wild environments here and there we chose not to lay waste. Your challenge is to create new kinds of plants and animals by genetic engineering and somehow fit them together into free-living artificial ecosystems. We understand that this feat may prove impossi-

ble. We are certain that for many of you even the thought of doing so will be repugnant. We wish you luck. And if you go ahead and succeed in the attempt, we regret that what you manufacture can never be as satisfying as the original creation. Accept our apologies and this audiovisual library that illustrates the wondrous world that used to be.

CHAPTER 4

———o———

THE PLANETARY KILLER

One of the most memorable events of my life occurred on a late May evening in 1994, in a back room of the Cincinnati Zoo, when I walked up to a four-year-old Sumatran rhinoceros named Emi, gazed into her lugubrious face for a while, and placed the flat of my hand against her hairy flank. She made no response except maybe to blink her eyes. That's it; that's all that happened. No matter: I had at last met my real-life unicorn.

The Sumatran rhino, *Dicerorhinus sumatrensis,* is a very special animal. Shy and elusive in the extreme, it is one of the rarest species in the world, classified as "critically endangered" in the Red List of the IUCN – World Conservation Union. Probably fewer than four hundred existed the evening I met Emi, and the number has continued to drop. Now, as I write in 2001, only about three hundred survive, of which seventeen are in captivity. The species may have no more than a few decades to live. At least one expert, Thomas Foose, gives it less than a fifty-fifty chance to reach mid-century.

The Sumatran rhino is a legend among wildlife and conservation biologists. Those who search for it in its native forest rarely get so much as a glimpse of a flesh-and-blood animal. Usually the most they can hope for are wallows and footprints along riverbanks and ridge tops or, slightly better, a rustle in the undergrowth and hint of musk in the air. I will never enjoy even that experience. Instead I have my memory of Emi, and a tuft of Sumatran rhino hair I keep on my desk as a talisman of *Dicerorhinus sumatrensis* and with it all the other vanishing species of the world.

The Sumatran is also special because it is a living fossil. The genus to which it belongs originated in the Oligocene Epoch at least thirty million years ago—halfway back to the Age of Dinosaurs—making it the oldest relatively unchanged mammal lineage in the world except for several ancient tropical bats. I could not help thinking that evening what an extraordinary and frightening time we live in, that I could see and touch this marvel from the other side of the world in what may be the final geological instant of its existence.

My host was the director of the Cincinnati Zoo, Edward Maruska, a fellow Sumatran rhino devotee. He explained that three adults had been assembled there, and others were being sought, as part of the international effort to build a backup breeding population against the possibility that the species goes extinct in the wild. At night the world of the captives consisted of wet concrete and steel bars for security. In the day the Sumatrans moved out to adjoining enclosures for public exhibition, a stroll through simulated natural conditions, and a feast of fifty kilos of fodder. The nocturnal compound I entered was filled with continuously playing soft-rock music. The melodies served to habituate the animals to sound and make them less likely to panic in case of a sudden noise—the slam of a door, the flyover of an airplane.

Rhinoceroses were once among the rulers of Earth. Through tens of millions of years prior to the birth of humanity, a wide array of rhino species, from small hippolike forms to towering giants larger than elephants, were among the dominant large herbivores in forests and grasslands across most of the landmasses of the world. The Sumatran is one of five species still surviving. It is the only two-horned species in Asia. The Javan rhino, which is even rarer than the Sumatran, has a single horn. So does the Javan's close relative, the greater Indian rhino, third-largest land animal in the world (after the African and Asian elephants) and still numerous enough, with 2,500 individuals in the wild, to be officially classified by conservation biologists as merely "endangered" as opposed to "critically endangered." The black and the white rhinoceros, both confined to sub-Saharan Africa, have double horns like the Sumatran rhino but are otherwise very different from it and each other. They too are endangered, flirting with critical status.

The Sumatran is anatomically overall the most distinctive of the five living species. Although the smallest, around 1,000 kilograms at maturity, it is still huge relative to other animals. It is further distinguished by a trait present in the distant rhinocerotid ancestors but not in the other four living species: a coat of shaggy hair, short, crisp, and black in the newborn, long, reddish brown, and somewhat flexuous in young adults, and finally sparse, dark, and bristly in the old. A hairy rhinoceros is a hard image to get used to, but only because the Sumatran is so seldom seen alive or illustrated in natural-history texts.

Sumatran rhinos are specialized for life in hilly rainforest with abundant standing water. Powerful and agile climbers, they can crash through undergrowth and race up and down steep slopes when pursued. They also swim easily across streams and lakes, and a few have even been seen dog-paddling offshore in the open sea. During the day they lounge and roll about in ponds and mud wallows, cooling themselves and nudging on mud coats for protection from the vicious tabanid horseflies that swarm in Asia's lowland forests. At night they browse the understory of mature woodland and the more succulent bounty of saplings and bushes in treefall gaps and along riverbanks, while trampling vegetation and breaking off low tree branches for additional food with their short, stumpy horns. They wear trails from their wallows to prime feeding sites. They also visit salt licks, a mineral resource on which their lives depend. Being herbivores, they are unaggressive unless provoked: they fight only to defend themselves and their young or to drive off other rhinos that invade their territories.

Except for sporadic encounters to mate and the mothering by females of their young, Sumatrans are entirely solitary. In normal circumstances, which hardly exist anymore, each adult patrols a home range of ten to thirty square kilometers, changing its wallow sites and foraging trails whenever local forage is used up. Females give birth to single calves and keep them close by for three years. Then they drive the young away to search for their own home ranges. The record for longevity of a Sumatran rhino in captivity is forty-seven years. Because of the pressure of intense hunting, however, old age is probably rarely attained nowadays in the wild.

The decline of the Sumatran rhino has been gradual and subtle, not abrupt and catastrophic, more like morbidity from cancer than collapse from a heart attack. The pattern is typical of vanishing species. Within historical times Sumatrans ranged across a great sweep of forested terrain from India through Myanmar (formerly Burma) and Vietnam, thence south through peninsular Malaysia to Sumatra and Borneo. A million years or more ago they must have been familiar to the small-brained human ancestor *Homo erectus* as it spread into tropical southeastern Asia from the western and central parts of the continent. Perhaps these earliest people hunted Sumatrans, but their crude tools and the impenetrability of the woodland fastness of the rhinos would have made captures rare. Because of the Sumatrans' exceptional elusiveness and the wildness of their habitats, they remained moderately abundant over much of their range even into historical times. Around the salt licks in the Gunung Leuser National Park of northern Sumatra, as many as fourteen individuals were once counted within a square kilometer.

By the mid-1980s, such concentrations had almost disappeared. The entire population was down to between five hundred and nine hundred individuals, including sixteen in captivity. The northern race comprised as few as six or seven individuals, then limited to Myanmar. Other local populations consisted of one hundred individuals in peninsular Malaysia, thirty to fifty on Borneo, and four hundred to seven hundred on Sumatra. At the present time the decline continues unabated. The Burmese race is evidently gone, and the Bornean population is likely to follow soon. Total extinction of the species in the wild seems inevitable within several decades unless drastic measures reverse the trend.

Is the Sumatran rhino dying of old age? Has its time come, like Great Aunt Clarissa on her deathbed, and should we therefore just let it slip away?

No, absolutely not, ever. Banish the thought! The premise of such a notion is demonstrably and dangerously false. The Sumatran and every other species that disappears typically dies young, at least in a physiological sense. The idea that species pass through natural life stages is based on an erroneous analogy. An endangered species is not like a dying patient whose care is too expensive and futile to prolong. The opposite is true. The great majority of rare and declining species are composed of young, healthy individuals. They just need the room and time to grow and reproduce that human activity has denied them.

The principle is well illustrated by the California condor, *Gymnogyps californianus.* One of the world's largest flying birds, the condor disappeared from almost all of its range across North America and came to the edge of extinction, not because its heredity deteriorated, but because people destroyed most of its natural habitat and shot and poisoned the survivors. When only a dozen individuals remained in the wild, biologists captured and placed them with a previously confined breeding colony near San Diego. Given protection and uncontaminated food, the combined group flourished. A few individuals have recently been released in the Grand Canyon and other restricted parts of the original range. For at least a while longer—and we most prayerfully hope, for thousands of years to come—the California condor is again a free-living species. If its habitat were somehow restored across its prehistoric breeding range and the population left unmolested, it might soar on its nine-foot wings from Atlantic to Pacific. That will not happen any time soon, of course, if ever, but at least the American fauna has regained one of its most dramatic species.

Other eleventh-hour rescues from obliteration have confirmed the generally innate resilience of endangered species. The most dramatic was that of the Mauritian kestrel. This small hawk, found only on the Indian Ocean island of Mauritius, was down to a single breeding pair in 1974. Most conservationists had given it up for lost. But the heroic efforts of a bird breeder, Carl Jones, and his collaborators brought the population back. Now there are about two hundred pairs, some in captivity and others returned to the wild, perhaps half the number that were present before Mauritius was settled by humans. The near-death experience forced the population through a narrow bottleneck and stripped the kestrel of most of its earlier genetic variability, but fortunately there were not enough defects among the surviving genes to impair survival or fertility.

Because extreme rescue measures are expensive and time-consuming, they can be used to help only a minute fraction of the thousands of critically endangered species of plants and animals. These fortunate few are likely to be the large, the beautiful, and the charismatic. And not all such attempts using captive populations will succeed. Unfortunately, the prospects of the Sumatran rhinoceros in particular are unpromising. The species is considered the most difficult large mammal in the world to breed, exceeding even the giant panda. The obstacles include the very brief periods of ovulation by the females, the likely need of a male to induce ovulation, and the aggressively solitary nature of potential mates at all other times. Of the seventeen individuals in zoos and fenced-in rainforest enclosures, only three males and five females have mated. Among the females, only Emi in the Cincinnati Zoo has been pregnant. After repeated spontaneous abortions, and to the joy of conservationists, she bore a healthy male calf on September 13, 2001.

The causes of the decline of the Sumatran rhino in the wild are well understood, but so far unstoppable. The once trackless forests of tropical Asia are being logged for timber at a steepening pace, to be replaced with farms and oil-palm plantations. The decimation of its habitat, however, would not alone be necessarily fatal to the Sumatran. There still exist on Sumatra, Borneo, and peninsular Malaysia natural reserves large enough to sustain small but viable populations. Instead, the fatal pressure is poaching, which is intense enough all by itself to wipe out the species within a few years if it is not curbed. Driving it is the ravenous demand of Chinese traditional medicine, whose practitioners believe—on no firm evidence—that rhino horn cures a wide range of afflictions, from fever to laryngitis and lower back pain. The result has been a free-market death spiral for the Sumatran rhino. As rhino horn grows scarcer, prices soar, with the result that poaching increases and rhino horn becomes even scarcer and more expensive. In 1998, horn from the black rhino of Africa fetched up to $12,000 per kilo in Taipei, roughly the price of gold, while a kilo from the greater Indian rhino brought an astronomical $45,000. I have not learned the cost of Sumatran horn, but it is likely to be in the same range as the greater Indian, because in general the Chinese prefer Asian species over African.

The illegal slaughter of all the rhinos quickened during the 1970s as a bizarre and unintended consequence of the OPEC oil embargo. As oil prices climbed, so did per-capita income across much of the Arab world. Among those who benefited were young men from the impoverished nation of Yemen, who emigrated to the oil fields of Saudi Arabia in search of fortune. They were able to afford more expensive jambiyas—ceremonial daggers bestowed to celebrate the Yemeni rite of passage. Because the most esteemed jambiyas have handles of rhino horn, poaching intensified.

Abnormal levels of poaching for Chinese medicine and jambiya handles combined have devastated rhino populations everywhere as never before dreamed possible. In 1909–10, when Theodore Roosevelt led his African expedition inland from Mombasa, black rhinoceroses numbered about one million. America's great conservation president felt no moral compunction in shooting a few. In 1970 the black rhino population still held at 65,000, but due to the jambiya craze it crashed—to 15,000 by 1980 and 4,800 by 1985. Fifteen years later, only 2,400 black rhinos remained. In 1997 Yemen finally became a party to the Convention on International Trade in Endangered Species of Wild Fauna and Flora (CITES), and the influx of rhino horn from that source may be abating. But the demand in Asia for horn used in folk medicine stays as unrelenting as ever, and deadly enough to wipe out the Sumatran rhino.

Little wonder the pressure is intense: a poacher who can earn ten years' income with a single rhino kill is willing to risk imprisonment and even his own life. Unfortunately for the Sumatran rhino, the danger to the killer is not very great in tropical Asia's deep forests, where animals can be hunted silently and out of sight. In earlier times, when rhino horn was moderately priced, native hunters tracked Sumatrans whenever they found fresh spoor, but in an opportunistic fashion, making no special effort to single the species out. With prices soaring, casual hunters have become specialized predators, crisscrossing the forests in search of rhino signs. They trap the animals in camouflaged pits or skewer them with spike logs suspended above trails and dropped by trip wires. Then they dispatch the helpless animals with rifle shots, butcher them for meat, and remove their horns for transport to waiting middlemen. The end of this tragedy can be easily envisioned: just four hundred campsite feasts and five million dollars in retail horn sales mark the path to final extinction for the Sumatran rhino.

In September 1992 Alan Rabinowitz, an expert on Asian large mammals, led an expedition to the Danum Valley of Sabah, at the northern end of Borneo, to search for some of the last of the Sumatran rhinos. The Danum Valley, which is maintained as a wildlife-protected area, is believed to harbor the largest number of the great island's dwindling population. The search group split into five teams, three entering the forest on foot and two putting down near its center by helicopter. Each team then worked its way across and back out of the valley by different routes. Altogether they found evidence of at most seven animals. They encountered long-abandoned wallows and "ghost spoor," furrows dug in the wallows by the horns of rhinos now presumed dead. They also discovered poachers. By accident one of the helicopter teams landed almost on top of a camp of the hunters, who scattered and fled.

Afterward, Rabinowitz and his colleague George Schaller visited Tamanthi, a sanctuary set up in Myanmar twenty years previously for the protection of tigers, Sumatran rhinos, and other large native mammals. Tigers were still present in small numbers, but there was no sign at all of rhinos. Local Lisu hunters described how they had tracked the animals down one by one until all were gone. The same was true for a second Burmese population that once lived near the northern limit of the species' range. Gone, said the hunters, none seen for many years. A few older men among them recalled the day the last one was killed, butchered, and dehorned.

Can the Sumatran rhino, like the California condor and the Mauritian kestrel, be pulled back from the grave? Of the two standard methods, captive breeding has so far been unproductive, while protection against poachers in existing reserves remains tenuous at best. The small circle of rhino experts working on the problem agree that *Dicerorhinus sumatrensis* has entered the endgame. Whatever the solution, they say, it must be found now or never. One new approach is to create sanctuaries intermediate in size between the zoo and the wild reserve by fencing in large, carefully monitored sections of rainforest. Constructions of this kind, working at a scale of a hundred acres or so, are under way in Sumatra, peninsular Malaysia, and Sabah. So far the enclosures have failed to yield baby rhinos, but at least their conditions are seminatural and, one hopes, conducive to reproduction. Meanwhile, given the madness of it all—exorbitant prices of rhino horn, the lack of scientific evidence for its curative powers, and the awful environmental price its continued use imposes—the best hope is that somehow the practitioners of traditional Chinese medicine can be persuaded or, if necessary, coerced to drop rhino horn from their pharmacopoeia.

Moral righteousness in the Western industrialized countries in such matters, while easily felt, is nonetheless far from justified. The same unbridled market forces are at work there as elsewhere in the world. For five hundred years weavers in the Kashmiri city of Srinagar have worked with wool of the Tibetan antelope, which is of fine enough quality as to have earned the Persian name *shahtoosh,* "wool of kings." In the late 1980s, *shahtoosh* shawls became an international craze, worn innocently, for example, by Queen Elizabeth II and the supermodel Christie Brinkley. The demand drove the annual shawl production from a few hundred items a year to thousands. The price of a single shawl rose to as high as $17,000. The Tibetan antelope was, of course, relentlessly hunted for more and more wool. It takes the hides of three or more antelopes to make a six-foot shawl, and today, with the *shahtoosh* trade still legal in Kashmir, an estimated 20,000 of the animals are killed each year. Only about 75,000 survive in the wild, mostly in the remote western and north-central parts of the Tibetan Plateau.

In the United States the strong demand for abalone along the California coast resulted in a decline of the four shallow-water species that are commercially harvested. (I was one of the unthinking consumers.) The shortage shifted effort to the white abalone, the deep-water and less accessible species, which also happens to be the most tender and desirable species. There followed an intense white-abalone fishery between 1969 and 1977, at the end of which the species had been reduced to endangered status. Today, devastated still further by poaching, the white abalone has all but completely disappeared.

The Sumatran rhino and white abalone are textbook examples of the legion of species around the world so savaged by overharvesting and other human activities as to be one short step from what conservation scientists call "global" extinction, leaving no surviving individuals anywhere. Animals in the most extreme peril belong to what I call the Hundred Heartbeat Club, species consisting of one hundred or fewer individuals, hence that number of heartbeats away from global extinction. Notable members include the Philippine eagle, the Hawaiian crow, the Spix's macaw, the Chinese river dolphin, the Javan rhino, the Hainan gibbon, the Vancouver Island marmot, the Texas pipefish, and the Indian Ocean coelacanth. Among many others lined up for early admission are the giant panda, the mountain gorilla, the Sumatran orangutan, the Sumatran rhino, the golden bamboo lemur, the Mediterranean monk seal, the Philippine crocodile, and the barndoor skate, the latter one of the largest fishes of the North Atlantic.

At least 976 tree species out of the 100,000 known to science worldwide are in similar difficulty. In one extreme category—so close to the edge that conservationists call them the "living dead"—are three species, including the Chinese hornbeam *(Carpinus putoensis)*, each of which survives as a single individual. Three others, including the beautiful *Hibiscus clayi* of Hawaii, are down to three or four individuals. The record for concentrated endangerment in one place may be held by the Juan Fernández Islands, a tiny archipelago six hundred kilometers off the coast of Chile and famous as the hermitage of Alexander Selkirk, fictionalized as Robinson Crusoe in Daniel Defoe's 1719 novel. On the seventy square miles of land live 125 species of plants found nowhere else. Because of the centuries of damage from visitors and residents and the fire, axes, and goats they brought, twenty of the species are represented by 25 or fewer individual plants in the wild. Six are small trees in the Juan Fernández endemic genus *Dendroseris*. One of the species, *Dendroseris macracantha,* was thought to survive as a single tree growing in a local garden. In the 1980s this individual was inadvertently cut down, and the species given up for lost—until one more survivor was discovered by a local tourist guide on a steep volcanic ridge in the interior. A species of sandalwood peculiar to Juan Fernández is believed to be extinct, although hope persists that one or two surviving trees may yet turn up.

It should come as no surprise that large numbers of species are crossing the thin zone from the critically endangered to the living dead and thence into oblivion. Yet a few authors—none of them biologists—still doubt that extinction is occurring on a large scale. Perhaps they have been misled by the circumstance that the extinction of a species, like the death of a human being, is seldom directly observed at the instant of its occurrence. It is further true that, because of their extreme rarity, endangered species are hard to locate in the first place. Statistically they remain in that parlous state only for a short time. In any given habitat on any day there are relatively few species in the red zone of the critically endangered. Many more threatened species are classified as merely "endangered" or, somewhat more comfortably, "vulnerable." The explanation is the same for the relative scarcity of patients in the intensive care ward of a hospital: just a little zig instead of a zag and they are gone.

Many recent extinctions are undoubtedly overlooked altogether, because the species were so rare they vanished before they were discovered and given a name. A near miss of this kind, famous among conservation biologists, is the po'ouli of Hawaii, a warbler-sized bird so distinctive in anatomy that it is placed in a genus by itself, *Melamprosops*. Known only from fossils for a time, it was thought to have been extinguished long before the American colonists arrived. Then in the early 1970s a small population of living birds was found in an isolated valley forest. Twenty years later, however, the po'ouli had become so scarce even in this last redoubt that intensive searches turned up only a few isolated individuals. It is likely to disappear soon, if it is not already extinct, this time for real and forevermore. In other groups of animals less well known than birds, including the myriad fungi, insects, and fishes, this scenario is likely to be occurring thousands of times over, leaving no record to show that the vanished species ever existed in the first place.

It is within the best-studied groups of animals and plants that the magnitude of the slaughter can be most easily observed and measured. Sixteen of the 263 mammals native to Australia, for example, are known to have vanished since the arrival of the European settlers. Here is the toll, with the date each species was last seen: Darling Downs hopping mouse (1840s); white-footed tree rat (1840s); big-eared hopping mouse (1843); broad-faced potoroo (1875); eastern hare-wallaby (1890); short-tailed hopping mouse (1894); Alice Springs mouse (1895); long-tailed hopping mouse (1901); pig-tailed bandicoot (1920); todache wallaby (1927); desert bandicoot (1931); lesser bilby (1931); central hare-wallaby (1931); lesser stick-nest rat (1933); thylacine, or Tasmanian wolf (1933); and crescent nailtail wallaby (1964). It is highly probable that at least some of the rarer and less conspicuous Australian species still present in the early nineteenth century disappeared before they came to the attention of naturalists. Furthermore, in 1996 an additional 34 species—14 percent of the entire surviving Australian mammal fauna—were red-listed by the IUCN – World Conservation Union as either vulnerable, endangered, or critically endangered.

Mass extinctions in Australia did not begin with the arrival of Western civilization. The cataclysm of its mammals during the past two centuries is only the latest episode in a much longer history of the decline of the overall fauna. Sixty thousand years ago, before the arrival of the aboriginal people, the island continent was home to many unique land animals of exceptionally large size. There was an abundance of giant flightless birds *(Genyornis newtoni)* related to the modern emus, but shorter-legged and, at 80 to 100 kilograms (180 to 220 pounds), twice their weight. Also present, and probably preying on the *Genyornis* emus and their egg clutches, were monitor lizards similar to the present-day Komodo dragon of Indonesia, but truly dinosaurian in size, reaching

seven meters (twenty-three feet) in length. They lived among creatures vaguely resembling giant sloths, rhinoceroses, and lions, as
well as oversized kangaroos and a horned terrestrial tortoise the
size of a small automobile. This megafauna must have been millions of years old, but it came to an abrupt end about the time the
first aboriginal people arrived. These oldest known colonists
crossed over to Australia from present-day Indonesia between
53,000 and 60,000 years ago. Not long afterward, and evidently
no later than 40,000 years ago, the megafauna had vanished. Not
a single land-dwelling species larger than a human being survived.
Also extinct were many other mammals, reptiles, and flightless
birds between one and fifty kilograms in weight. Biologists, by
dating eggshell fragments of *Genyornis* emus with the use of iso-
tope ratios, have determined that the species disappeared from all
over Australia in a short period of time about 50,000 years ago. Its
extinction cannot be easily ascribed to climatic change, or disease,
or volcanic activity. The disappearance of the mammoth bird,
however, does correlate approximately with the arrival of the first
human beings. It appears that the European colonists of Australia,
long afterward and aided by their companion rats, rabbits, and
foxes, have merely carried the extinction process to the next level
beyond that inflicted by the aboriginals.

Humanity, when wiping out biodiversity, eats its way down the
food chain. First to go among animal species are the big, the slow,
and the tasty. As a rule around the world, wherever people entered
a virgin environment, most of the megafauna soon vanished. Also
doomed were a substantial fraction of the most easily captured
ground birds and tortoises. Smaller and swifter species were able
to hang on in diminished numbers.

Archaeologists have found that the extinction spasms occur
within a period of a few centuries or, at most, millennia following
the arrival of colonists. The history of the fauna of Madagascar is
a textbook example. This great island off the coast of Africa was
isolated at least eighty-eight million years ago, at the time it broke
free from India. Afterward the two landmasses continued sepa-

rately in a northward drift toward Asia. During this time Madagascar evolved dramatically distinct life forms. Two thousand
years ago, before voyagers from Indonesia first arrived, it was an
untouched zoological garden of big animals. Its forest and grasslands harbored tortoises with carapaces as much as four feet
across; cow-sized pygmy hippos; a mongoose the size of a lynx;
and the Malagasy "aardvark," so unusual in anatomy that zoologists have placed it in a special order of its own, the Bibymalagasia.
Also present were a half dozen kinds of elephant birds, ranging
from ostrich-size up to the largest of the *Aepyornis* species, which
stood ten feet tall, weighed half a ton, and laid eggs the size of soc-
cer balls. Arab traders working the north Madagascar coast around
the ninth century knew of these great birds from either oral tradition or personal experience recounted by contemporary Malagasy.
The birds then made their way into the legends of *The Thousand
and One Nights* as the *rukh*—the rocs—eaglelike behemoths that
could seize and carry away elephants. Equally myth-worthy were
the lemurs, which are primitive primates and hence very distant
cousins of human beings. The fifty or so lemur species of primeval
Madagascar included giants such as a 60-pound arboreal apelike
form that traveled upside-down along branches like a tree sloth; a
120-pound ecological equivalent of Australia's arboreal, leaf-eating
koala; and a ground-dwelling species larger than an adult male
gorilla that probably filled the same niche as the extinct ground
sloths of the New World.

The oldest known archaeological site on Madagascar as I write
dates to A.D. 700. By the eleventh century the island was extensively occupied by agricultural villages and cattle-herding
settlements. Simultaneously—and this can hardly be a coinci-
dence—virtually all of the native species of mammals, birds, and
reptiles above 10 kilograms (roughly, 20 pounds) vanished. The
single exception was the wily and far-ranging Nile crocodile.
Native stories suggest that one or two of the big lemurs may have
survived in remote forests until the seventeenth century, but
carbon-dated remains that recent are still lacking. No fewer than

fifteen species of lemurs, one-third of the total fauna, disappeared when the human tide washed over Madagascar. All the animals that vanished had been diurnal, as well as physically larger than those alive today, and as a result especially vulnerable to the Malagasy colonists. The case against humans for the extinction of the megafauna is built solely on circumstantial evidence, but the facts would win at least an indictment in any court of law.

The trail of *Homo sapiens,* serial killer of the biosphere, reaches to the farthest corners of the world. A few centuries after the Madagascan slaughter, another occurred in New Zealand. When the Polynesians came ashore in the late thirteenth century they, like the Malagasy colonists, entered a vast biological wonderland. The most conspicuous animals were the moas, large flightless birds resembling ostriches and emus but independently evolved on these islands alone. The smallest of the eleven known moa species were about the size of turkeys. The largest, the prodigious *Dinornis giganteus* ("gigantic terrible-bird") stood nine feet tall and weighed over 150 kilograms (330 pounds). Because New Zealand is remote from Australia and other landmasses, it lacked native mammals. Across millions of years moas had filled their niches. The birds served environmentally as woodchucks, rabbits, deer, and rhinoceroses rolled into one phylogenetic group of close kin. They were further diversely specialized for life in major habitats of the islands, from mountain to lowland and wet forest to dry scrub and grassland. In their prehuman domain they had only one known predator, the giant New Zealand eagle, *Harpagornis moorei,* which weighed about 13 kilograms (28 pounds). Then, like the sweep of a scythe, came the Maori. Spreading from north to south, they butchered the moas in huge numbers and piled their bones conspicuously in hunting sites all over the islands. By the middle of the fourteenth century, a matter of a few decades, the moas were gone and presumably the world's largest eagle with them.

The suddenness of the mass extinction has been documented in painful detail by recent archaeological studies. Possibly no more than a hundred colonists first arrived, and as few as a thousand were present when the last of the moas disappeared. Yet this small human population was enough to destroy an estimated 160,000 moas. The birds, being flightless and probably tame, were easily caught. Usually only the upper legs were eaten, and the remainder of the carcasses were discarded or fed to the camp dogs. Almost certainly the eggs were consumed whenever nests were found. The moas could not keep pace with such high-intensity predation, even when the human carnivores were few. The reproductive rate of the great birds was just too low: only one or two eggs were laid in a clutch, and chicks required as long as five years to mature. Mathematical models allow that small bands of humans could easily have wiped out an entire fauna with such a demographic handicap.

The damage of the occupation of New Zealand then grew more complex and bit deeper. Rats inadvertently introduced by the colonists multiplied to become a deadly enemy of the unprotected smaller birds, reptiles, and amphibians. As their numbers grew, the Maoris cleared and burned extensive areas, reducing some of the habitats to marginal levels. A total of twenty other land birds, including eight flightless species, were eliminated. In the 1800s the newly arrived British settlers swept their own lethal blade across New Zealand, introducing additional alien plants and animals and converting far more of the natural environments to cropland and pasture. Of the eighty-nine bird species known to be unique to New Zealand before the Maoris came, only fifty-three, or 60 percent, survive today.

The New Zealand event was only the final chapter in mass extinctions that began on islands to the north. What we celebrate in the colonization of Polynesia as a grand historical epic for humanity was for the rest of life a rolling wave of destruction. The vast triangle of archipelagoes that embrace the Pacific are a natural laboratory for the study of extinction. Only during the past twenty years has research revealed the full extent of the human impact on its living environment. The first settlers from the East

Indies penetrated Micronesia and parts of Melanesia about four thousand years ago. Then, sailing eastward from one island stepping stone to the next, their descendants colonized Fiji, Samoa, and Tonga thirty-five hundred years ago, the Marquesas two thousand years ago, and New Zealand, Hawaii, and Easter Island within the past fifteen hundred years. Pressing forward relentlessly, multiplying, filling all the habitable land, they exterminated about half of the bird species on each island group in turn. With the arrival of Europeans, bringing advanced agriculture, technology, diseases, and a demon swarm of ants, mosquitoes, weeds, and other invasive species, the destruction continued unabated. The two waves of invasion have eliminated or driven to extinction not only the core of the two thousand or so bird species that originally inhabited the Pacific archipelagoes, but also many other animals and plants as well.

The extinction of species is a worldwide phenomenon, and it extends beyond animals killed for food to the plants and countless small animals that depend upon them. This progression of megadeath has followed the filter principle of conservation biology: the farther back in time the first human-induced wave of extinction struck, the lower the extinction rate today. For example, fewer species are now endangered in Samoa, Tonga, and other parts of the western Pacific, the first to be colonized by humans, than in Hawaii, the most recently colonized of the major island groups. The explanation is elementary and in no way comforting. The most vulnerable species, such as tortoises and large ground birds, were the first in Polynesia to go. When these "wimps" were no longer present, the more resilient native animals became targets and hung on for varying periods afterward. When the fauna of a given island has been reduced in size, only the most resistant species are left to endanger. On the island of Eua, in Tonga, for example, the number of forest birds has dropped from twenty-seven or more in prehuman times (three thousand years ago and earlier) to ten by the end of the nineteenth century, and finally to nine today.

The progress of filtering has been documented in precise detail by archaeologists at Shag River Mouth of New Zealand's South Island, the site of one of the most active moa-hunter base camps. The strata of bone deposits reveal that the hunters started with the largest moas and the equally accessible seals and penguins. When these became scarce, the settlers shifted to smaller moas, dogs, songbirds, fish, and mussels and other shellfish. Within several decades of their arrival at Shag River Mouth, all the moas were evidently gone, and the hunters had left.

As a consequence of the filtering effect, the decline of biodiversity is most difficult to chart in those parts of the world where human occupancy is oldest. Nevertheless, painstaking research can sometimes restore the historical essentials, especially when both human activity and the old faunas and floras are tracked through time together. The lands bordering the northern and eastern Mediterranean, for example, have been occupied almost since the origin of modern *Homo sapiens* and *Homo neanderthalensis*, our extinct sibling human species. For hundreds of thousands of years, these populations were very sparse and widely scattered. Their refuse middens, now fossilized, reveal that they depended heavily on easily collected tortoises and marine shellfish such as mussels and oysters. Toward the end of the Ice Age, ten thousand years ago, with the Neanderthals by then long gone and replaced by Cro-Magnon *Homo sapiens* across ice-free Europe, *sapiens* populations in western Asia invented agriculture and began to convert large swaths of wild land into cultivated fields. Supported by wheat, goats, and other domesticated species, and hence able to live in population densities ten to a hundred times those of the Cro-Magnon hunter-gatherers, the agriculturists spread west and north at an average rate of one kilometer per year. Within four thousand years their farms and villages reached from the Fertile Crescent to England. As the slow-breeding and vulnerable tortoises grew scarce, the human populations turned increasingly to fast-breeding but more elusive partridges, rabbits, and hare. At roughly the same time, much of the megafauna of Europe disappeared, including the woolly mammoth, the woolly rhinoceros (a relative of the Sumatran rhino), the cave bear, the Cyprus pygmy hippopotamus, and the gigantic deer known as the Irish elk.

All of the continents except Antarctica supported megafaunas of one kind or another before the spread of humanity. Only Africa and tropical Asia were spared the full shock of extinction. How are we to explain this anomaly? The answer appears to be that, for hundreds of millennia, evolving humanity was a native species itself in Africa and Asia. The two continents, united geographically as almost a single supercontinent, are the cradle of the human species. They have been mostly occupied since the ancestral *Homo erectus* advanced across them more than a million years ago. Early *Homo sapiens* emerged, spread, and coevolved there with the rest of the native fauna and flora, which as a result had time to adapt genetically to its presence. During this period humans shared predators and diseases with the other animals, preyed upon them, and served as prey in turn. They were too scarce and technologically primitive to pose a great threat to the rest of biodiversity. In sharp contrast, the modern races of *Homo sapiens* were a true alien species when they colonized the rest of the world, from Australia to the New World and finally the distant oceanic islands. Like the rats, pigs, and assorted diseases they carried with them, they met few coevolved prey and enemies. Adapting to the new environments by culture at a rate thousands of times faster than possible with genes alone, they outpaced any defense that the resident biotas could raise. So down went the native organisms.

The biotas continue to fall before our remorseless expansion, in ever-rising numbers across ever-broader arrays of plants and animals. Where originally it was mostly large land-dwelling animals that were afflicted, now fishes, amphibians, insects, and plants are, for the first time, vanishing in large numbers. The dawnless night of extinction is also descending upon rivers, lakes, estuaries, coral reefs, and even the open sea.

How much extinction is occurring today? Researchers generally agree that it is catastrophically high, somewhere between one thousand and ten thousand times the rate before human beings began to exert a significant pressure on the environment. The previous natural, or Edenic, period of biodiversity, as measured by paleontologists, started with the beginning of the Phanerozoic Era 450 million years ago. It ended fifty to ten millennia ago, with the ascent of Upper Paleolithic and Neolithic peoples, whose improved tools, dense populations, and deadly efficiency in the pursuit of wildlife inaugurated the current extinction spasm.

In Edenic times extinction rates averaged, to the nearest order of magnitude (power of ten), one species per million each year. Of course, occasional episodes of mass extinction punctuated the half-billion years, and there were occasional long stretches of relative tranquillity. These natural catastrophes were followed by episodes of very rapid evolution, as surviving species multiplied to fill the niches opened to them. One part of the world differed from another according to climatic circumstance. Extinction rates also varied among the groups of organisms, from a known high of one species per half-million in mammals to a known low of one species per six million in echinoderms. However, when the rates are averaged over all the groups best documented in the fossil record and across tens of millions of years, the grand average of extinction rates comes out to be very roughly—nearest order of magnitude—one species per million species each year.

The Edenic rate of new species formation was also about one species per million per year, again to the nearest order of magnitude, approximately balancing extinction. Actually, the species birthrate was slightly higher than the species death rate, allowing the standing global number to grow very slowly through geological time. Biodiversity, counted by species, or genera, or families, and on both land and sea, is today about twice as great as that averaged over the past 450 million years.

Because of the importance of extinction, I will now explain briefly the several ways used by biologists to estimate its current rates and why the methods produce different numbers. First, if we simply count the number of extinctions actually observed during the past century in well-studied "focal" groups such as birds and flowering plants, the annual rate is only ten to a hundred per million. But this is much too low, because the causes of extinction intensified throughout the twentieth century. They are now the highest ever, and still rising. Further, large numbers of species, while not erased to the last individual, are nevertheless so rare and declining so rapidly as to be "committed" to extinction in the near future. And not least, numerous species undoubtedly were so rare and local so as to escape discovery before they vanished, and hence are not on the books.

So in a second step, let us assume that all of the species in the IUCN – World Conservation Union Red List, the endangered species census of record, are destined to fall to extinction or be committed to early extinction within the next hundred years. For example, the 2000 Red List estimates that almost one in four of Earth's mammal species and one in eight of the bird species are at some degree of risk. Now we are no longer counting extinctions in the past but instead commitments to extinction in the near future. As a result, the estimated annual rate of extinction jumps to between one hundred and one thousand species per million. But again this must be too low, because as the causes of extinction intensify, more and more new species are pouring into the threatened categories of the Red Lists and sliding down the ratchet toward oblivion. When that acceleration is taken into account, the rate jumps to between one thousand and ten thousand.

Three independent measures have been used to arrive at the final and higher range of estimates—in other words, one thousand to ten thousand species per million per year. Despite their relative crudity, the methods are persuasively consistent. The first and most commonly employed measure, already described in the last chapter, is the relation between the area of a habitat and the number of species it can support. As a forest, grassland, or stream system is reduced, the number of species it can support over a period of years drops in a predictable manner. In almost all cases it drops at between the third and sixth root of the area left to it.

The second method is to track the descent of species through the Red Lists over a period of years. Many pass from either secure or unknown status to vulnerable to endangered to critically endangered and finally, after many fruitless searches, are judged extinct. Extremely few reverse direction and return to a safer category. The velocity of flow by masses of species through the Red List categories yields an estimate of the number that will pass to extinction in the future.

The third method, based on ecological knowledge, is to analyze the survival probabilities of the species in the different Red List categories. The likelihood that a threatened species will live or die depends on the size of its populations, how widely they are dispersed and exchange individuals, how much they fluctuate through time, and the longevity and reproductive rate of the organisms composing them. This is the technique called Population Viability Analysis, or PVA for short. Although still only a weak contributor to the study of whole faunas and floras, the method is being improved rapidly by biologists and is certain to play a major role in the future of conservation forecasts.

All sound empirical science consists of successive approximations that employ multiple approaches and trial-and-error measurements. These approximations inform and are themselves informed by increasingly sophisticated theory. The estimation of extinction rates is typical of this reciprocity between fact and theory. In time the procedures used to analyze extinction will grow more precise than the current spread of order-of-magnitude approximations.

Although it is possible to predict species extinction for the near future—say, over the next decade or two—such a projection is impossible for the more distant future. The obvious reason is that the trajectory depends on human choice. If the decision were taken today to freeze all conservation efforts at their current level while allowing the same rates of deforestation and other forms of environmental destruction to continue, it is safe to say that at least a fifth of the species of plants and animals would be gone or committed to early extinction by 2030, and half by the end of the century. If, on the other hand, an all-out effort is made to save the biologically richest parts of the natural world, the amount of loss can be cut by at least half.

The somber archaeology of vanished species has taught us the following lessons:

- The noble savage never existed.
- Eden occupied was a slaughterhouse.
- Paradise found is paradise lost.

Humanity has so far played the role of planetary killer, concerned only with its own short-term survival. We have cut much of the heart out of biodiversity. The conservation ethic, whether expressed as taboo, totemism, or science, has generally come too late and too little to save the most vulnerable of life forms.

If Emi the Sumatran rhino could speak, she might tell us that the twenty-first century is thus far no exception. And I would respond with another reassuring touch of my hand. We know more about the problem now, Emi; it is not too late. We know what to do. Perhaps we will act in time.

CHAPTER 5

———o———

HOW MUCH IS THE
BIOSPHERE WORTH?

In the early nineteenth century the coastal plain of the southern United States was much the same as in countless millennia past. From Florida and Virginia west to the Big Thicket of Texas, primeval stands of cypress and flatland hardwoods wound around the corridors of longleaf pine through which the early Spanish explorers found their way into the continental interior. The signature bird of this wilderness, a dweller of the deep bottomland woods, was the ivory-billed woodpecker, *Campephilus principalis.* Its large size, exceeding a crow's, its flashing white primaries, visible at rest, and its loud nasal call, *kent!* . . . *kent!* . . . *kent!,* likened by Audubon to the false high note of a clarinet, made it both conspicuous and instantly recognizable. Mated pairs worked together up and down the boles and through the canopies of high trees, clinging to vertical surfaces with splayed claws while hammering their powerful off-white beaks through dead wood into the burrows of beetle larvae and other insect prey. The hesitant beat of their strikes, *tick tick* . . . *tick tick tick* . . . *tick tick* . . . heralded their approach from a distance in the dark woods. They came to the observer like spirits out of an unfathomed wilderness core.

Alexander Wilson, early American naturalist and friend of Audubon, assigned the ivorybill noble rank. Its manners, he wrote in *American Ornithology* (1808–14), "have a dignity in them superior to the common herd of woodpeckers. Trees, shrubbery, orchards, rails, fence posts, and old prostrate logs, are all alike interesting to those, in their humble and indefatigable search for prey; but the royal hunter before us, scorns the humility of such situations, and seeks the most towering trees of the forest; seeming particularly attached to those prodigious cypress swamps whose crowded giant sons stretch their bare and blasted or moss-hung arms midway to the sky."

A hundred years later, almost all of the virgin bottomland forest had been replaced by farms, towns, and second-growth woodlots. Shorn of its habitat, the ivorybill declined precipitously in numbers. By the 1930s it was down to scattered pairs in the few remaining primeval swamps of South Carolina, Florida, and Louisiana. In the 1940s the only verifiable sightings were in the Singer Tract of northern Louisiana. Afterward, only rumors of sightings persisted, and even those faded with each passing year.

The final descent of the ivorybill was closely watched by Roger Tory Peterson, whose classic *A Field Guide to the Birds* had fired my own interest in birds as a teenager. In 1995, the year before he died, I met Peterson, one of my heroes, for the first and only time. I asked him a question common in conversations among American naturalists. What of the ivory-billed woodpecker? He gave the answer I expected: "Gone."

I thought, surely not gone *everywhere,* not globally! Naturalists are among the most hopeful of people. They require the equivalent of an autopsy report, cremation, and three witnesses before they write a species off, and even then they would hunt for it in séances if they thought there was any chance of at least a virtual image. Maybe, they speculate, there are a few ivorybills in some inaccessible cove, or deep inside a forgotten swamp, known only to a few close-mouthed cognoscenti. In fact, several individuals of a small Cuban race of ivorybills were discovered during the 1960s in an isolated pine forest of Oriente Province. Their current status is unknown. In 1996 the Red List of the IUCN – World Conservation Union reported the species to be everywhere extinct, including Cuba. I have heard of no further sightings, but evidently no one at this writing knows for sure.

Why should we care about *Campephilus principalis?* It is, after all, only one of ten thousand bird species in the world. Let me give a simple and I hope decisive answer: because we knew this particular species, and knew it well. For reasons difficult to understand and express, it became part of our culture, part of the rich mental world of Alexander Wilson and all those afterward who cared about it. There is no way to make a full and final valuation of the ivorybill or any other species in the natural world. The measures we use increase in number and magnitude with no predictable limit. They rise from scattered unconnected facts and elusive emotions that break through the surface of the subconscious mind, occasionally to be captured by words, although never adequately.

We, *Homo sapiens,* have arrived and marked our territory well. Winners of the Darwinian lottery, bulge-headed paragons of organic evolution, industrious bipedal apes with opposable thumbs, we are chipping away the ivorybills and other miracles around us. As habitats shrink, species decline wholesale in range and abundance. They slip down the Red List ratchet, and the vast majority depart without special notice. Being distracted and self-absorbed, as is our nature, we have not yet fully understood what we are doing. But future generations, with endless time to reflect, will understand it all, and in painful detail. As awareness grows, so will their sense of loss. There will be thousands of ivory-billed woodpeckers to think about in the centuries and millennia to come.

Is there any way now to measure even approximately what is being lost? Any attempt is almost certain to produce an underestimate, but let me start anyway with macroeconomics. In 1997 an international team of economists and environmental scientists put a dollar amount on all the ecosystems services provided humanity free of charge by the living natural environment. Drawing from multiple databases, they estimated the contribution to be $33 trillion or more each year. This amount is nearly twice the 1997 combined gross national product (GNP) of all countries in the world, or gross world product, of $18 trillion. Ecosystems services are defined as the flow of materials, energy, and information from the biosphere that support human existence. They include the regulation of the atmosphere and climate; the purification and retention of fresh water; the formation and enrichment of the soil; nutrient cycling; the detoxification and recirculation of waste; the pollination of crops; and the production of lumber, fodder, and biomass fuel.

The 1997 megaestimate can be expressed in another, even more cogent, manner. If humanity were to try to replace the free services of the natural economy with substitutes of its own manufacture, the global GNP would have to be raised by at least $33 trillion. The exercise, however, cannot be performed except as a thought experiment. To supplant natural ecosystems entirely, even mostly, is an economic and even physical impossibility, and we would certainly die if we tried. The reason, ecological economists explain, is that the marginal value, defined as the rate of change in the value of ecosystems services relative to the rate of fall in the availability of these services, rises sharply with every increment in the fall. If taken too far, the rise will outpace human capacity to sustain the needed services by combined natural and artificial means. Hence, a much greater dependence on artificial means—in other words, environmental prostheses—puts at risk not just the biosphere but also humanity itself.

Most environmental scientists believe that the shift has already been taken too far, lending credit to the folk injunction "Don't mess with Mother Nature." The lady is our mother all right, and a mighty dispensational force as well. After evolving on her own for more than three billion years, she gave birth to us a mere million years ago, an eye blink in evolutionary time. Ancient and vulnerable, she will not tolerate the undisciplined appetite of her gargantuan infant much longer.

Abundant signs of the biosphere's limited resilience exist all around. The oceanic fish catch now yields $2.5 billion to the U.S. economy and $82 billion worldwide. But it will not grow further, simply because the amount of ocean is fixed and the organisms it can generate is static. As a result, all of the world's seventeen oceanic fisheries are at or below sustainable yield. During the 1990s the annual global catch leveled off at about 90 million tons. Pressed by ever-growing global demand, it can be expected eventually to drop. Already fisheries of the western North Atlantic, the Black Sea, and portions of the Caribbean have collapsed. Aquaculture, or the farming of fish, crustaceans, and mollusks, takes up part of the slack, but at rising environmental cost. This "fin-and-shell revolution" necessitates the conversion of valuable wetland habitats, which are nurseries for marine life. To feed the captive populations, fodder must be diverted from crop production. Thus aquaculture competes with other human activity for productive land while reducing natural habitat. What was once free for the taking must now be manufactured. The ultimate result will be an upward inflationary pressure across wide swaths of the world's coastal and inland economies.

Another case in point: forested watersheds capture rainwater and purify it before returning it by gradual runoffs to the lakes and sea, all for free. They can be replaced only at great cost. For generations New York City thrived on exceptionally clean water from the Catskill Mountains. The watershed inhabitants were proud that their bottled water was once sold throughout the Northeast. As their population grew, however, they converted more and more of the watershed forest into farms, homes, and resorts. Gradually the sewage and agricultural runoff adulterated the water until it fell below Environmental Protection Agency standards. Officials in New York City now faced a choice. They could build a filtration plant to replace the Catskill Watershed, at

$6 billion to $8 billion capital cost, followed by $300 million annual running costs; or else they could restore the Catskills Watershed to somewhere near its original purification capacity for $1 billion, with subsequently very low maintenance costs. The decision was easy, even for those born and bred in an urban environment. In 1997 the city raised an environmental bond issue and set out to purchase forested land and to subsidize the upgrading of septic tanks in the Catskills. There is no reason the people of New York City and the Catskills cannot enjoy the double gift from nature in perpetuity of clean water at low cost and a beautiful recreational area at no cost.

There is even a bonus in the deal. By its policy of natural water management, the Catskill forest region also secures flood control at very little expense. The same benefit is available to the city of Atlanta. When 20 percent of the trees in the metropolitan area were removed during its rapid development, the result was an annual increase in stormwater runoff of 4.4 billion cubic feet. If Atlanta were to build enough containment facilities to contain this volume, the cost would be at least $2 billion. In contrast, trees replanted along streets, yards, and available parking space are a great deal cheaper than concrete drains and revetments. Their maintenance cost is near zero, and, not least, they are more pleasing to the eye.

In conserving nature, whether for practical or aesthetic reasons, diversity matters. The following rule is now widely accepted by ecologists: the more species that inhabit an ecosystem, such as a forest or lake, the more productive and stable is the ecosystem. By "production," the scientists mean the amount of plant and animal tissue created each hour or year or any other given unit of time. By "stability" they mean one or the other or both of two things: first, how narrowly the summed abundances of all species vary through time; and second, how quickly the ecosystem recovers from fire, drought, and other stresses that perturb it. Human beings understandably wish to live in the midst of diverse, productive, and stable ecosystems. Who, if given a choice, would build their home in a wheat field instead of a parkland?

Ecosystems are kept stable in part by the insurance principle of biodiversity. If a species disappears from a community, its niche will be more quickly and effectively filled by another species if there are many candidates for the role instead of few. Example: A ground fire sweeps through a pine forest, killing many of the understory plants and animals. If the forest is biodiverse, it recovers its original composition and production of plants and animals more quickly. The larger pines escape with some scorching of their lower bark and continue to grow and cast shade as before. A few kinds of shrubs and herbaceous plants also hang on and resume regeneration immediately. In some pine forests subject to frequent fires, the heat of the fire itself triggers the germination of dormant seeds genetically adapted to respond to heat, speeding the regrowth of forest vegetation still more.

A second example of the insurance principle: When we scan a lake, our macroscopic eye sees only relatively big organisms—eelgrass, pondweeds, fishes, waterbirds, dragonflies, whirligig beetles, and other things big enough to splash and go bump in the night. But all around them, in vastly greater numbers and variety, are the invisible bacteria, protistans, planktonic single-celled algae, aquatic fungi, and other microorganisms. These seething myriads are the true foundation of the lake ecosystem and the hidden agents of its stability. They decompose the bodies of the larger organisms. They form large reservoirs of carbon and nitrogen, release carbon dioxide, and thereby damp fluctuations in the organic cycles and energy flows in the rest of the aquatic ecosystem. They hold the lake close to a chemical equilibrium, and, to a point, they pull it back from extreme perturbations caused by silting and pollution.

In the dynamism of a healthy ecosystem there are minor players and major players. Among the major players are the ecosystems engineers, which add new parts to the habitat and open the door to guilds of organisms specialized to use them. Biodiversity grows more biodiversity, and the overall abundance of plants, animals, and microorganisms increases to a corresponding degree:

- By constructing dams, beavers create ponds, bogs, and flooded meadows. These environments shelter species of plants and animals that are rare or absent in free-running streams. The submerged masses of decaying wood forming the dams add still more species that occupy and feed on them.
- Elephants trample and tear up shrubs and small trees, opening glades within forests. The result is a mosaic of habitats containing overall larger numbers of resident species.
- Florida gopher tortoises dig thirty-foot-long tunnels that mix and diversify the texture of the soil, altering the composition of its microorganisms. Their retreats are also shared by snakes, frogs, and ants specialized to live in the burrows.
- *Euchondrus* snails of Israel's Negev Desert grind down soft rocks to feed on the lichens growing inside. By converting rock to soil and releasing the nutrients photosynthesized by the lichens, they multiply niches for other species.

Overall, a large number of independent observations from differing kinds of ecosystems point to the same conclusion: the more species that live together, the more stable and productive the ecosystems they compose. On the other hand, mathematical models that attempt to describe the interactions of species in ecosystems show that the seemingly opposite also occurs: high diversity can lower the stability of individual species. Under certain conditions, including random colonization of the ecosystem by large numbers of species that interact strongly with one another, the separate but interlocking fluctuations of species can increase volatility in species abundances, causing species to vary more widely in abundance and hence rendering them more liable to extinction. Similarly, given appropriate species traits, it is mathematically possible for increased diversity to lead to lower production.

When observation and theory collide, scientists turn to carefully designed experiments for resolution. Their motivation is especially high in the case of biological systems, which are typically far too complex to be grasped by observation and theory alone. The best procedure, as in the rest of science, is first to simplify the system, then to hold it more or less constant while varying the important parameters one or two at a time to see what happens. In the 1990s a team of British ecologists, in an attempt to approach these ideal conditions, devised the ecotron, a growth chamber in which artificially simple ecosystems can be assembled as desired, species by species. Using multiple ecotrons, they found that productivity, measured by the increase of plant bulk, rose with an increase in species numbers. Simultaneously, ecologists monitoring patches of Minnesota grassland—outdoor equivalents of ecotrons—during a period of drought found that patches higher in species diversity declined less in productivity and recovered more quickly than did patches with lower diversity.

These pioneering experiments appeared to uphold the conclusion drawn earlier from natural history, at least with reference to production. Put more precisely, ecosystems tested thus far do not possess the qualities and starting conditions allowed by theory that can lower production and produce instability as a result of large species numbers.

But—how can we be sure, the critics asked (pressing on in the best tradition of science), that the increase in production in particular is truly the result of just an increase in the number of species? Maybe the effect is due to some other factor that just happens to be correlated with species numbers. Perhaps the result is a statistical artifact. For example, the larger the number of plant species present in a habitat, the more likely it is that at least one kind among them will be extremely productive. If that occurs, the rise in yield of plant tissue—and the animals feeding on it—is only a matter of luck of the draw and not the result of some pure property of biodiversity itself. At its base the distinction made by this alternative hypothesis is semantic. The increased likelihood of acquiring an outstandingly productive species can be viewed as just one means by which the enrichment of biodiversity raises productivity. (If you draw on a pool of one thousand candidates for a basketball team, you are more likely to get a star than if you draw on a pool of one hundred candidates.) Still, it is important to know if other consequences of biodiversity enrichment play important roles. In particular, do species interact in a manner that increases the growth of either one or both? This is the process called "overyielding." In the mid-1990s a massive study was undertaken to test the effect of biodiversity on productivity that paid special attention to the presence or absence of overyielding. Multiple projects of BIODEPTH, as the project came to be called, were conducted during a two-year period by thirty-four researchers in eight European countries. This time the results were more persuasive. They showed once again that productivity does rise with biodiversity, at least in groupings of up to thirty-two species. Many of the experimental runs also revealed the existence of overyielding.

Over millions of years, nature's ecosystems engineers have been especially effective in the promotion of overyielding. They have coevolved with other species that exploit the niches they build. The result is a harmony within ecosystems. The constituent species, by spreading out into multiple niches, seize and cycle more materials and energy than is possible in similar ecosystems. *Homo sapiens* is an ecosystems engineer too, but a bad one. Not having coevolved with the majority of life forms we now encounter around the world, we eliminate far more niches than we create. We drive species and ecosystems into extinction at a far higher rate than existed before and everywhere diminish productivity and stability.

I will grant at once that economic and production values at the ecosystem level do not alone justify saving every species in an ecosystem, especially those so rare as to be endangered. The loss of the ivory-billed woodpecker has had no discernible effect on American prosperity. A rare flower or moss could vanish from the

Catskill forest without diminishing the region's filtration capacity. But so what? To evaluate individual species solely by their known practical value at the present time is business accounting in the service of barbarism. In 1973 the economist Colin W. Clark made this point persuasively in the case of the blue whale, *Balaenopterus musculus*. At 100 feet and 150 tons in full maturity, the species is the largest animal that ever lived on the land or in the sea. It is also among the easiest to hunt and kill. Over 300,000 were harvested during the twentieth century, with a peak haul of 29,649 in the 1930–31 season. By the early 1970s the population had plummeted to several hundred individuals. The Japanese were especially eager to continue the hunt even at the risk of total extinction. So Clark asked, What practice would yield the whalers and humanity the most money: cease hunting and let the blue whales recover in numbers and then harvest them sustainably forever, or kill the rest off as quickly as possible and invest the profits in growth stocks? The disconcerting answer for annual discount rates over 21 percent: kill them all and invest the money.

Now, let us ask, what is wrong with that argument?

Colin Clark's implicit answer is simple. The dollars-and-cents value of a dead blue whale was based only on the measures relevant to the existing market—that is, on the going price per unit weight of whale oil and meat. There are many other values, destined to grow along with our knowledge of living *Balaenopterus musculus* in science, medicine, and aesthetics, in dimensions and magnitudes still unforeseen. What was the value of the blue whale in A.D. 1000? Close to zero. What will be its value in A.D. 3000? Essentially limitless, plus the gratitude of the generation then alive to those who, in their wisdom, saved the whale from extinction.

No one can guess the full future value of any kind of animal, plant, or microorganism. Its potential is spread across a spectrum of known and as yet unimagined human needs. Even the species themselves are largely unknown. Fewer than two million are in the scientific register, with formal Latinized names, while an estimated five million to one hundred million—or more—await discovery. Of the species known, fewer than 1 percent have been studied beyond the sketchy anatomical descriptions used to diagnose them.

Agriculture is one of the vital industries most likely to be upgraded by attention to the remaining wild species. The world's food supply hangs by a slender thread of biodiversity. Ninety percent is provided by slightly more than a hundred plant species out of a quarter-million known to exist. Twenty species carry most of the load, of which only three—wheat, maize, and rice—stand between humanity and starvation. For the most part the premier twenty are those that happened to be present in the regions where agriculture was independently invented some ten thousand years ago, namely the Mediterranean perimeter and Near East; Central Asia; the horn of Africa; the rice belt of tropical Asia; and the uplands of Mexico, Central America, and Andean South America. Yet some thirty thousand species of wild plants, most occurring outside these regions, have edible parts consumed at one time or other by hunter-gatherers. Of these, at least ten thousand can be adapted as domestic crops. A few, including the three species of New World amaranths, the carrotlike arracacha of the Andes, and the winged bean of tropical Asia, are immediately available for commercial development.

In a more general sense, all the quarter-million plant species—in fact, all species of organisms—are potential donors of genes that can be transferred by genetic engineering into crop species in order to improve their performance. With the insertion of the right snippets of DNA, new strains can be created that are variously cold-hardy, pest-proofed, perennial, fast-growing, highly nutritious, multipurpose, water-conservative, and more easily sowed and harvested. And compared with traditional breeding techniques, genetic engineering is all but instantaneous.

The method, a spin-off of the revolution in molecular genetics, was first developed in the 1970s. During the 1980s and 1990s, before the world quite realized what was happening, it came of age. A gene from the bacterium *Bacillus thuringiensis,* for example, was inserted into the chromosomes of corn, cotton, and potato plants, allowing them to manufacture a toxin that kills insect pests. No need to spray insecticides: the engineered plants now perform this task on their own. Other bacterial transgenes, as they are called, were inserted into soybean and canola plants to make them resistant to chemical weed killers. Agricultural fields can now be cheaply cleared of weeds with no harm to the crops growing there. The most important advance of all, achieved in the late 1990s, was the creation of golden rice. This new strain is laced with bacterial and daffodil genes that allow it to manufacture beta-carotene, a precursor of vitamin A. Because rice, the principal food of three billion people, is deficient in vitamin A, the addition of beta-carotene is no mean humanitarian feat. About the same time, the almost endless potential of genetic engineering was confirmed by two circus tricks of the trade: a bacterial gene was implanted into a monkey and a jellyfish bioluminescence gene into a plant.

It was nevertheless inevitable that genetic engineering would be less than universally dazzling, and stir opposition. For many, human existence was being transformed in a fundamental and insidious way. With little warning, genetically modified organisms (GMOs) had entered our lives and were all around us, incomprehensibly changing the order of nature and society. Protest movements against the new industry began in the mid-1990s and exploded in 1999, just in time to rank as a millennial event with apocalyptic overtones. The European Union banned transgenic crops, the Prince of Wales compared the methodology to playing God, and radical activists called for a global embargo of all GMOs. "Frankenfoods," "superweeds," and "Farmageddon" entered the vocabulary: they were, according to one British newspaper, the "mad forces of genetic darkness." Some prominent environmental scientists found technical and ethical reasons for concern.

As I write (in 2001), public opinion and official policy have come to vary greatly from one country to the next. France and

Britain are vehemently opposed. China is strongly favorable, and Brazil, India, Japan, and the United States cautiously so. In the United States particularly, the public awoke to the issue only after the transgenie (so to speak) was out of the bottle. From 1996 to 1999 the amount of U.S. farmland devoted to genetically modified crops had rocketed from 3.8 million to 70.9 million acres. As the century ended, more than half of soybean and cotton grown was engineered, as well as nearly a third (28 percent) of corn.

There are, in fact, several sound reasons for anxiety over genetic engineering, which I will now summarize and evaluate.

• Many people, not just philosophers and theologians, are troubled by the ethics of transgenic evolution. They grant the benefits but are unsettled by the reconstruction of organisms in bits and pieces. Of course, human beings have been creating new strains of plants and animals since agriculture began, but never at the sweep and pace inaugurated by genetic engineering. And during the era of traditional plant breeding, hybridization was used to mix genes almost always among varieties of the same species or, at most, closely similar species. Now it is across entire kingdoms, from bacteria and viruses to plants and animals. How far the process should be allowed to continue remains an open ethical issue.

• The effects on human health of each new transgenic food are hard to predict, and certainly never free of risk. However, the product can be tested just like any other new food product on the market, then certified and labeled. There is no reason at this time to assume that their effects differ in any fundamental way. Yet scientists generally agree that a high level of alertness is essential, and for the following reason: all genes, whether original to the organism or donated to it by an exotic species, have multiple effects. Primary effects, such as the manufacture of a pesticide, are the ones sought. But destructive secondary effects, including activity as allergens or carcinogens, are also at least a remote possibility.

• Transgenes can escape from the modified crops into wild relatives of the crop where the two grow close together. Hybridization has always occurred widely in agriculture, even before the advent of genetic engineering. It has been recorded at one time and place or another in twelve of the thirteen most important crops used worldwide. However, the hybrids have not overwhelmed their wild parents. I know of no case in which a hybrid strain outcompetes wild strains of the same or closely related species in the natural environment. Nor has any hybrid turned into a superweed, in the same class as the worst wild, nonhybrid weeds that afflict the planet. As a rule, domesticated species and strains are less competitive than their wild counterparts in both natural and human-modified environments. Of course, transgenes could change the picture. It is simply too early to tell.

• Genetically modified crops can diminish biological diversity in other ways. In a now famous example, the bacterial toxin used to protect corn is carried in pollen by wind currents for distances of sixty meters or more from the cultivated fields. Then, landing on milkweed plants, they are capable of killing the caterpillars of monarch butterflies feeding there. In another twist, when cultivated fields are cleared of weeds with chemical sprays to which the crops are protected by transgenes, the food supply of birds is reduced and their local populations decline. These environmental secondary effects have not been well studied in the field. How severe they will become as genetic engineering spreads remains to be seen.

• Many people, having become aware of the potential threats of genetic engineering in their food supply, understandably believe that yet another bit of their freedom has been taken from them by faceless corporations (who can even name, say, three of the key players?), using technology beyond their control or even understanding. They also fear that an industrialized agriculture dependent on high technology can by one random error go terribly wrong. At the heart of the anxiety is a sense of helplessness. In the realm of public opinion, genetic engineering is to agriculture as nuclear engineering is to energy.

The problem before us is how to feed billions of new mouths over the next several decades and save the rest of life at the same time, without being trapped in a Faustian bargain that threatens freedom and security. No one knows the exact solution to this dilemma. Most scientists and economists who have studied both sides of it agree that the benefits outweigh the risks. The benefits must come from an evergreen revolution. The aim of this new thrust is to lift food production well above the level attained by the green revolution of the 1960s, using technology and regulatory policy more advanced and even safer than those now in existence.

Genetic engineering will almost certainly play an important role in the evergreen revolution. Energized by recognition of both its promise and its risk, most countries have begun to fashion policies to regulate the marketing of transgenic crops. The ultimate driving force in this rapidly evolving process is international trade. In an important first step to address the issue, taken in 2000, over 130 countries tentatively agreed to the Cartagena Protocol on Biosafety, which provides the right to block imports of transgenic products. The protocol also sets up a joint "biosafety clearing house" to publish information on national policy. About the same time the U.S. National Academy of Sciences, joined by the science academies of five other countries—Brazil, China, India, Mexico, and the United Kingdom—as well as the Third World Academy of Sciences, endorsed the development of transgenic crops. They made recommendations for risk assessment and licensing agreements and stressed the needs of the developing countries in future research programs and capital investment.

Medicine is another domain that stands to gain enormously from the world's store of biodiversity, with or without the impetus of genetic engineering. Pharmaceuticals in current use are already drawn heavily from wild species. In the United States about one-quarter of all prescriptions dispensed by pharmacies are substances extracted from plants. Another 13 percent originate from micro-organisms and 3 percent more from animals, for a total of about 40 percent. Even more impressively, nine of the ten leading prescription drugs originally came from organisms. The commercial value of the relatively small number of natural products is substantial. The over-the-counter cost of drugs from plants alone was estimated in 1998 to be $20 billion in the United States and $84 billion worldwide.

In spite of its obvious potential, however, only a tiny fraction of biodiversity has been utilized in medicine. The narrowness of the base is illustrated by the dominance of ascomycete fungi in the control of bacterial diseases. Although only about thirty thousand species of ascomycetes have been studied, and they compose 2 percent of the total known species of organisms, they have yielded 85 percent of the antibiotics in current use. The underutilization is still greater than these figures alone suggest: probably fewer than 10 percent of the world's ascomycete species have even been discovered and given a scientific name. The flowering plants have been similarly scanted. Although it is likely that more than 80 percent of the species have received a scientific name, only some 3 percent of this fraction have been assayed for alkaloids, the class of natural products that have proved to be among the most potent curative agents for cancer and many other diseases.

There is an evolutionary logic in the pharmacological bounty of wild species. Throughout the history of life all kinds of organisms have evolved chemicals needed to control cancer in their own bodies, kill parasites, and fight off predators. Mutations and natural selection, which invent this armamentarium, are processes of endless trial and error. Hundreds of millions of species, evolving by the life and death of astronomical numbers of organisms across geological stretches of time, have yielded the present-day winners of the mutation-and-selection lottery. We have learned to consult them while assembling a large part of our own pharmacopoeia. Thus antibiotics, fungicides, antimalarial drugs, anesthetics, analgesics, blood thinners, blood-clotting agents, agents that prevent clotting, cardiac stimulants and regulators, immunosuppressive agents, hormone mimics, hormone inhibitors, anticancer drugs, fever suppressants, inflammation controls, contraceptives, diuretics and antidiuretics, antidepressants, muscle relaxants, rubefacients, anticongestants, sedatives, and abortifacients are now at our disposal, compliments of wild biodiversity.

Revolutionary new drugs have rarely been developed by the pure insights of molecular and cellular biology, even though these sciences have grown very sophisticated and address the causes of disease at the most fundamental level. Rather, the pathway of discovery has usually been the reverse: the presence of the drug is first detected in whole organisms, and the nature of its activity is subsequently tracked down to the molecular and cellular levels. Then the basic research begins.

The first hint of a new pharmaceutical may lie among the hundreds of remedies of Chinese traditional medicine. It may be spotted in the drug-laced rituals of an Amazonian shaman. It may come from a chance observation by a laboratory scientist unaware of its potential importance for medicine. More commonly nowadays, the clue is deliberately sought by the random screening of plant and animal tissues. If a positive response is obtained—say, a suppression of bacterial or cancer cells—the molecules responsible can be isolated and tested on a larger scale, using controlled experiments with animals and then (cautiously!) human volunteers. If the tests are successful, and the atomic structure of the molecule is also in hand, the substance can be synthesized in the laboratory, then commercially, usually at lower cost than by extraction from harvested raw materials. In the final step, the natural chemical compounds provide the prototype from which new classes of organic chemicals can be synthesized, adding or taking away atoms and double bonds here and there. A few of the novel substances may prove more efficient than the natural prototype. And of equal importance to the pharmaceutical companies, these analogs can be patented.

Serendipity is the hallmark of pharmacological research. A

chance discovery can lead not only to a successful drug but to advances in fundamental science, which in time yield other successful drugs. Routine screening, for example, revealed that an obscure fungus growing in the mountainous interior of Norway produces a powerful suppressor of the human immune system. When the molecule was isolated from the fungal tissue and identified, it proved to be a complex molecule of a kind never before encountered by organic chemists. Nor could its effect be explained by the contemporary principles of molecular and cellular biology. But its relevance to medicine was immediately obvious, because when organs are transplanted from one person to another, the immune system of the host must be prevented from rejecting the alien tissue. The new agent, named cyclosporin, became an essential part of the organ transplant industry. It also served to open new lines of research on the molecular events of the immune response itself.

The surprising events that sometimes lead from natural history to medical breakthrough would make excellent science fiction—if only they were untrue. The protagonists of one such plot are the poison dart frogs of Central and South America, belonging to the genera *Dendrobates* and *Phyllobates* in the family Dendrobatidae. Tiny, able to perch on a human fingernail, they are favored as terrarium animals for their beautiful colors: the forty known species are covered by various patterns of orange, red, yellow, green, or blue, usually on a black background. In their natural habitat dendrobatids hop about slowly and are relatively unfazed by the approach of potential predators. For the trained naturalist their lethargy triggers an alarm, in observance of the following rule of animal behavior: if a small and otherwise unknown animal encountered in the wild is strikingly beautiful, it is probably poisonous; and if it is not only beautiful but also easy to catch, it is probably deadly.

And so it is with dendrobatid frogs, which, it turns out, secrete a powerful toxin from glands on their backs. The potency varies according to species. A single individual of one (perfectly named)

Colombian species, *Phyllobates horribilis,* for example, carries enough of the substance to kill ten men. The Indians of two tribes living in the Andean Pacific slope forests of western Colombia, the Emberá Chocó and the Noanamá Chocó, rub the tips of their blowgun darts over the backs of the frogs, very carefully, then release the little creatures unharmed so they can make more poison. In the 1970s a chemist, John W. Daly, and a herpetologist, Charles W. Myers, gathered material from a similar Ecuadoran frog, *Epipedobates tricolor,* for a closer look at the dendrobatid toxin. In the laboratory Daly found that very small amounts administered to mice worked as an opiumlike painkiller, yet otherwise lacked the properties of typical opiates. Would it also prove nonaddictive? If so, the substance might be turned into the ideal anesthetic. From a cocktail of compounds taken from the backs of the frogs, Daly and his fellow chemists isolated and characterized the toxin itself, a molecule resembling nicotine, which they named epibatidine. This natural product proved two hundred times more effective in the suppression of pain than opium, but it was unfortunately also too toxic for practical use. The next step was to redesign the molecule. Chemists at Abbott Laboratories synthesized not only epibatidine, but hundreds of novel molecules resembling it. When tested clinically, one of the products, code-named ABT-594, was found to combine the desired properties. It depresses pain like epibatidine, including that from nerve damage of a kind usually impervious to opiates, and it is also nonaddictive. And ABT-594 has two other advantages: it promotes alertness instead of sleepiness and has no side effects on respiration or digestion.

The full story of the poison dart frogs also carries a warning about the conservation of tropical forests. The discovery of epibatidine and its synthetic analogs almost never occurred, thanks to the destruction of the habitat in which populations of *Epipedobates* live. By the time John Daly and Charles Myers set out to collect enough toxin for chemical analysis, following their initial visit to Ecuador, one of the two prime rainforest sites occupied by the frogs had been cleared and replaced with banana plantations. At the second site, which fortunately was still intact, they were able to find enough individuals to harvest just one milligram of the poison. From that tiny sample chemists were able, with skill and luck, to identify epibatidine and launch a major new initiative in pharmacological research.

It is no exaggeration to say that the search for natural medicinals is a race between science and extinction, and will become critically so as more forests fall and coral reefs bleach out and disintegrate. Another adventure dramatizing this point started in 1987, when the botanist John Burley collected samples of plants from a swamp forest near Lundu in the Malaysian state of Sarawak, on the northwestern corner of the island of Borneo. His expedition was one of many launched by the National Cancer Institute (NCI) to search for new natural substances to add to the fight against cancer and AIDS. Following routine procedure, the team collected a kilogram of fruit, leaves, and twigs from each kind of plant they encountered. Part was sent to the NCI laboratory for assay, and part was deposited in the Harvard University Herbarium for future identification and botanical research.

One such sample came from a small tree at Lundu about twenty-five feet high. It was given the voucher code label Burley-and-Lee 351. Back at the NCI laboratories, an extract made from it was tested routinely against human cancer cells grown in culture. Like the majority of such preparations, it had no effect. Then it was run through screens designed to test its potency against the AIDS virus. The NCI scientists were startled to observe that Burley-and-Lee 351 gives, in their words, "100 percent protection against the cytopathic effects of HIV-1 infection," having "essentially halted HIV-1 replication." In other words, while the substance the sample contained does not cure AIDS, it can stop cold the development of disease symptoms in HIV-positive patients.

The Burley-and-Lee 351 tree was determined to be a species of *Calophyllum*, a group of species belonging to the mangosteen family, or Guttiferae. Collectors were dispatched to Lundu a second time to obtain more material from the same tree, with the aim of isolating and chemically identifying the HIV inhibitor. The tree was gone, probably cut down by local people for fuel or building materials. The collectors returned home with samples taken from other *Calophyllum* trees in the same swamp forest, but their extracts were ineffective against the virus.

Peter Stevens, then at Harvard University, and the world authority on *Calophyllum,* now stepped in to solve the problem. The original tree, he found, belonged to a rare strain named *Calophyllum lanigerum* variety *austrocoriaceum.* The trees sampled on the second trip were another species, explaining their inactivity. No more specimens of *austrocoriaceum* could be found at Lundu. The search for the magic strain widened, and finally a few more specimens were located in the Singapore Botanic Garden. Now supplied with enough raw material, chemists and microbiologists were able to identify the anti-HIV substance as (+)-calanolide A. Soon afterward, the molecule was synthesized and proved as effective as the raw extract. Additional research revealed that it is a powerful inhibitor of reverse transcriptase, an enzyme needed by the HIV virus to replicate itself within the human host cell. Studies are now under way to determine the suitability of calanolide for market distribution.

The exploration of wild biodiversity in search of useful resources is called bioprospecting. Propelled by venture capital, it has in the past ten years grown into a respectable industry within a global market hungry for new pharmaceuticals. It is also a means for discovering new food sources, fibers, petroleum substitutes, and other products. Sometimes bioprospectors screen many species of organisms in search of chemicals with particular qualities, such as antisepsis or suppression of cancer. On other occasions bioprospecting is opportunistic, keying on one or a few species that show signs of yielding a valuable resource. Ultimately, entire ecosystems will be prospected as a whole, assaying all of the species for most or all of the products they can yield.

The extraction of wealth from an ecosystem can be destructive or benign. Dynamiting coral reefs and clearcutting forests yield fast profits but are unsustainable. Fishing coral reefs lightly and gathering wild fruit and resins in otherwise undisturbed forests are sustainable and long-lived. Collecting samples of valuable species from rich ecosystems and cultivating them in bulk elsewhere in biologically less-favored areas is not only profitable but the most sustainable of all.

Bioprospecting with minimal disturbance is the way of the future. Its promise can be envisioned with the following matrix for a hypothetical forest. To the left make a list of the thousands of plant, animal, and microbial species, as many as you can, recognizing that the vast majority have not yet been examined and that many still lack even a scientific name. Along the top prepare a horizontal row of the hundreds of functions imaginable for all the products of these species combined. The matrix itself is the combination of the two dimensions. The spaces filled within the matrix are the potential applications, whose nature remains almost wholly unknown.

The richness of biodiversity's bounty is reflected in the products already extracted by native peoples of the tropical forests, using local knowledge and low technology of a kind transmitted solely by demonstration and oral teaching. Here, for example, is a small selection of the most common medicinal plants used by tribes of the upper Amazon. Their knowledge has evolved from their combined experience with the more than fifty thousand species of flowering plants native to the region: motelo sanango, *Abuta grandifolia* (snakebite, fever); dye plant, *Arrabidaea chica* (anemia, conjunctivitis); monkey ladder, *Bauhinia guianensis* (amoebic dysentery); Spanish needles, *Bidens alba* (mouth sores, toothache); firewood tree, or capirona, species of *Calycophyllum* and *Capirona* (diabetes, fungal infection); wormseed, *Chenopodium ambrosioides* (worm infection); caimito, *Chrysophyllum cainito* (mouth sores, fungal infections); toad vine, *Cissus sicyoides* (tumors); renaquilla, *Clusia rosea* (rheumatism, bone fractures); calabash, *Crescentia cujete* (toothache); milk tree, *Couma macro-*

carpa (amoebic dysentery, skin inflammation); dragon's blood, *Croton lechleri* (hemorrhaging); fer-de-lance plant, *Dracontium loretense* (snakebite); swamp immortelle, *Erythrina fusca* (infections, malaria); wild mango, *Grias neuberthii* (tumors, dysentery); wild senna, *Senna reticulata* (bacterial infection).

Only a few of the thousands of such traditional medicinals used in tropical forests around the world have been tested by Western clinical methods. Even so, the most widely used already have commercial value that rivals farming and ranching. In 1992 a pair of economic botanists, Michael Balick and Robert Mendelsohn, demonstrated that single harvests of wild-grown medicinals from two tropical forest plots in Belize were worth $726 and $3,327 per hectare (2.5 acres) respectively, with labor costs thrown in. By comparison, other researchers estimated per-hectare yield from tropical forest converted to farmland at $228 in nearby Guatemala and $339 in Brazil. The most productive Brazilian plantations of tropical pine could yield $3,184 from a single harvest.

In short, medicinal products from otherwise undisturbed tropical forests can be locally profitable, providing markets are developed and the extraction rate is kept low enough to be sustainable. And when plant and animal food products, fibers, carbon credit trades, and ecotourism are added to the mix, the commercial value of sustainable use can be boosted far higher.

Examples of the new economy in practice are growing in number. In the Petén region of Guatemala, about six thousand families live comfortably by sustainable extraction of rainforest products. Their combined annual income is $4 million to $6 million, more than could be made by converting the forest into farms and cattle ranches. Ecotourism remains a promising but largely untapped additional resource.

Nature's pharmacopoeia has not gone unnoticed by industry strategists. They are well aware that even a single new molecule has the potential to recoup a large capital investment made in bioprospecting and product development. The single greatest success to date was achieved with extremophile bacteria living in the boiling-hot thermal springs of Yellowstone National Park. In 1983 Cetus Corporation used one of the organisms, *Thermus aquaticus,* to produce a heat-resistant enzyme needed for synthesis of DNA. The manufacturing process, called polymerase chain reaction (PCR), is today the foundation of rapid genetic mapping, a stanchion of the new molecular biology and medical genetics. By allowing microscopic amounts of DNA to be multiplied and typed, it also plays a key role in crime detection and forensic medicine. Cetus's patents on PCR technology, which have been upheld by the courts, are immensely profitable, with annual earnings now in excess of $200 million and growing.

Bioprospecting can serve both mainstream economics and conservation when done on a firm contractual basis. In 1991 Merck signed an agreement with Costa Rica's National Institute of Biodiversity (INBio) to assist in the search for new pharmaceuticals in Costa Rica's rainforests and other natural habitats. The first deposit was $1 million dispensed over two years, with two similar consecutive grants to follow. During the first period the field collectors concentrated on plants, in the second on insects, and in the third on microorganisms. Merck is now working through the immense library of materials it collected in this period, testing and refining chemical extracts made from them.

Also in 1991, Syntex signed a contract with Chinese science academies to receive up to ten thousand plant extracts a year for pharmaceutical assays. In 1998 Diversa Corporation signed on with Yellowstone National Park to continue bioprospecting the hot springs for biochemicals from thermophilic microbes. Diversa pays the park $20,000 yearly to collect the organisms for study, and a fraction of the profits generated by commercial development. Funds returning to Yellowstone will be used to promote conservation of the unique microbes and their habitat, as well as for basic scientific research and public education.

Still other agreements have been signed between NPS Pharmaceuticals and the government of Madagascar, between Pfizer and the New York Botanical Garden, and between the international company Glaxo Wellcome and a Brazilian pharmaceutical company, with part of the profits pledged to the support of Brazilian science.

Perhaps it is enough to argue that the preservation of the living world is necessary for our long-term material prosperity and health, as I have now done. But as I hope to show next, there is another, and in some ways deeper, reason. It has to do with the defining qualities and self-image of the human species.

CHAPTER 6

———o———

FOR THE LOVE OF LIFE

Have you ever wondered how we will be remembered a thousand years from now, when we are as remote as Charlemagne? Many would be satisfied with a list that includes the following: *the technoscientific revolution continued, globalized, and unstoppable; computer capacity approaching that of the human brain; robotic auxiliaries proliferating; cells rebuilt from molecules; space colonized; population growth slackening; the world democratized; international trade accelerated; people better fed and healthier than ever before; life span stretched; religion holding firm.*

In this buoyant vision of the twenty-first century, what might we have overlooked about our place in history? What are we neglecting and at risk of forever losing? The answer most likely in the year 3000 is: *much of the rest of life, and part of what it means to be a human being.*

A few technophiles, I expect, will beg to differ. What, after all, in the long term does it mean to be human? We have traveled this far; we will go on. As to the rest of life, they continue, we should be able to immerse fertilized eggs and clonable tissues of endangered species in liquid nitrogen and use them later to rebuild the destroyed ecosystems. Even that may not be necessary: in time entirely new species and ecosystems, better suited to human needs than the old ones, can be created by genetic engineering. *Homo sapiens* might choose to redesign itself along the way, the better to live in a new biological order of our own making.

Such is the extrapolated endpoint of technomania applied to the natural world. The compelling response, in my opinion, is that to travel even partway there would be a dangerous gamble, a single throw of the dice with the future of life on the table. To revive or synthesize the thousands of species needed—probably millions when the still largely unknown microorganisms have been cataloged—and put them together in functioning ecosystems is beyond even the theoretical imagination of existing science. Each species is adapted to particular physical and chemical environments within the habitat. Each species has evolved to fit together with certain other species in ways biologists are only beginning to understand. To synthesize ecosystems on bare ground or in empty water is no more practicable than the reanimation of deep-frozen human corpses. And to redesign the human genotype better to fit a ruined biosphere is the stuff of science horror fiction. Let us leave it there, in the realm of imagination.

Another reason exists not to take the gamble, not to let the natural world slip away. Suppose, for the sake of argument, that new species can be engineered and stable ecosystems built from them. With that distant potential in mind, should we go ahead, and for short-term gain, allow the original species and ecosystems to slip away? Yes? Erase Earth's living history? Then also burn the libraries and art galleries, make cordwood of the musical instruments, pulp the musical scores, erase Shakespeare, Beethoven, and Goethe, and the Beatles too, because all these—or at least fairly good substitutes—can be re-created.

The issue, like all great decisions, is moral. Science and technology are what we can do; morality is what we agree we should or should not do. The ethic from which moral decisions spring is a norm or standard of behavior in support of a value, and value in turn depends on purpose. Purpose, whether personal or global, whether urged by conscience or graven in sacred script, expresses the image we hold of ourselves and our society. In short, ethics evolve through discrete steps, from self-image to purpose to value to ethical precepts to moral reasoning.

A conservation ethic is that which aims to pass on to future generations the best part of the nonhuman world. To know this world is to gain a proprietary attachment to it. To know it well is to love and take responsibility for it.

Each species—American eagle, Sumatran rhinoceros, flat-spined three-toothed land snail, furbish lousewort, and on down the roster of ten million or more still with us—is a masterpiece. The craftsman who assembled them was natural selection, acting upon mutations and recombinations of genes, through vast numbers of steps over long periods of time. Each species, when examined closely, offers an endless bounty of knowledge and aesthetic pleasure. It is a living library. The number of genes prescribing a eukaryotic life form such as a Douglas fir or a human being runs into the tens of thousands. The nucleotide pairs composing them—in other words, the genetic letters that encode the life-giving enzymes—vary among species from one billion to ten billion. If the DNA helices in one cell of a mouse, a typical animal species, were placed end on end and magically enlarged to have the same width as wrapping string, they would extend for over nine hundred kilometers, with about four thousand nucleotide pairs packed into every meter. Measured in bits of pure information, the genome of a cell is comparable to all editions of the *Encyclopaedia Britannica* published since its inception in 1768.

The creature at your feet dismissed as a bug or a weed is a creation in and of itself. It has a name, a million-year history, and a place in the world. Its genome adapts it to a special niche in an ecosystem. The ethical value substantiated by close examination of its biology is that the life forms around us are too old, too complex, and potentially too useful to be carelessly discarded.

Biologists point to another ethically potent value: the genetic unity of life. All organisms have descended from the same distant ancestral life form. The reading of the genetic codes has shown thus far that the common ancestor of all living species was similar to present-day bacteria and archaeans, single-celled microbes with the simplest known anatomy and molecular composition. Because of this single ancestry, which arose on Earth over 3.5 billion years ago, all species today share certain fundamental molecular traits. Their tissue is divided into cells, whose enveloping lipid membranes regulate exchange with the outside environment. The molecular machinery that generates energy is similar. The genetic information is stored in DNA, transcribed into RNA, and translated into proteins. Finally, a large array of mostly similar protein catalysts, the enzymes, accelerate all the life processes.

Still another intensely felt value is stewardship, which appears to arise from emotions programmed in the very genes of human social behavior. Because all organisms have descended from a common ancestor, it is correct to say that the biosphere as a whole began to think when humanity was born. If the rest of life is the body, we are the mind. Thus, our place in nature, viewed from an ethical perspective, is to think about the creation and to protect the living planet.

As cognitive scientists have focused on the nature of the mind, they have come to characterize it not just as a physical entity, the brain at work, but more specifically as a flood of scenarios. Whether set in the past, present, or future, whether based on reality or entirely fictive, these free-running narratives are all churned out with equal facility. The present is constructed from the avalanche of sensations that pour into the wakened brain. Working at a furious pace, the brain summons memories to screen and make sense of the incoming chaos. Only a minute part of the information is selected for higher-order processing. From that part, small segments are enlisted through symbolic imagery to create the white-hot core of activity we call the conscious mind.

During the story-building process the past is reworked and then returned to storage. The repeated cycles allow the brain to hold on to only small but shrinking fragments of these former conscious states. Over a lifetime the details of real events are increasingly distorted by editing and supplementation. Across generations the most important among them turn into history, and finally legend and myth.

Each culture has its own creation myth, the primary functions of which are to place the tribe that contrived it at the center of the universe, and to portray history as a noble epic. The ultimate epic unfolding through science is the genetic history both of *Homo sapiens* and of all our antecedents. Traced back far enough through time, across more than three billion years, all organisms on Earth share a common ancestry. That genetic unity is a fact-based history confirmed with increasing exactitude by the geneticists and paleontologists who reconstruct evolutionary genealogy. If *Homo sapiens* as a whole must have a creation myth—and emotionally in the age of globalization it seems we must—none is more solid and unifying for the species than evolutionary history. That is another value favoring stewardship of the natural world.

To summarize: a sense of genetic unity, kinship, and deep history are among the values that bond us to the living environment. They are survival mechanisms for ourselves and our species. To conserve biological diversity is an investment in immortality.

Do other species therefore have inalienable rights? There are three reaches of altruism possible from which a response can be made. The first is anthropocentrism: nothing matters except that which affects humanity. Then pathocentrism: intrinsic rights should be extended to chimpanzees, dogs, and other intelligent animals for whom we can legitimately feel empathy. And finally biocentrism: all kinds of organisms have an intrinsic right at least to exist. The three levels are not as exclusive as they first seem. In real life they often coincide, and when in life-or-death conflict they can be ordered in priority as follows: first humanity, next intelligent animals, then other forms of life.

The influence of the biocentric view, expressed institutionally through quasi-religious movements such as Deep Ecology and the Epic of Evolution, is growing worldwide. The philosopher Holmes Rolston III tells a story that can serve as a parable of this trend. For years trailside signs at a subalpine campground in the Rocky Mountains he occasionally visited read, "Please leave the flowers for others to enjoy." When the wooden signs began to erode and flake, they were replaced by new ones that read, "Let the flowers live!"

It is not so difficult to love nonhuman life, if gifted with knowledge about it. The capacity, even the proneness to do so, may well be one of the human instincts. The phenomenon has been called biophilia, defined as the innate tendency to focus upon life and lifelike forms, and in some instances to affiliate with them emotionally. Human beings sharply distinguish the living from the inanimate. We esteem novelty and diversity in other organisms. We are thrilled by the prospect of unknown creatures, whether in the deep sea, the unbroken forest, or remote mountains. We are riveted by the idea of life on other planets. Dinosaurs are our icons of vanished biodiversity. More people visit zoos in the United States than attend professional sports events. Their favorite site in the National Zoo of Washington, D.C., is the insect exhibit, representing maximum novelty and diversity.

A prominent component of biophilia is habitat selection. Studies conducted in the relatively new field of environmental psychology during the past thirty years point consistently to the following conclusion: people prefer to be in natural environments, and especially in savanna or parklike habitats. They like a long depth of view across a relatively smooth, grassy ground surface dotted with trees and copses. They want to be near a body of water, whether ocean, lake, river, or stream. They try to place their habitations on a prominence, from which they can safely scan the savanna and watery environment. With nearly absolute consistency these landscapes are preferred over urban settings that are either bare or clothed in scant vegetation. To a relative degree people dislike woodland views that possess restricted depth of vision, a disordered complexity of vegetation, and rough ground structures—in short, forests with small, closely spaced trees and dense undergrowth. They want a topography and openings that improve their line of sight.

People prefer to look out over their ideal terrain from a secure position framed by the semienclosure of a domicile. Their choice of home and environs, if made freely, combines a balance of refuge for safety and a wide visual prospect for exploration and foraging. There may be small gender differences: among Western landscape painters at least, women stress refuges with small prospect spaces, and men stress large prospect spaces. Women also tend to place human figures in or near the refuges, while men place them more consistently in the open spaces beyond.

The ideal natural habitat is intuitively understood by landscape architects and real-estate entrepreneurs. Even when it offers no practical value, the setting commands a relatively high price, reaching its maximum if also located conveniently near cities.

I once described the principle of the ideal habitat to a wealthy friend as we looked down from his New York penthouse to the open woodland and lake of Central Park. His terrace, I also noticed, was ringed by potted plants. I thought of him as a convincing experimental subject. It has since often occurred to me that to see most clearly the manifestations of human instinct, it is useful to start with the rich, who among us enjoy the widest range of options in response, and most readily follow their emotional and aesthetic inclinations.

No direct evidence has yet been sought for a genetic basis of the human habitat preference, but its presence is suggested by a consistency in its manifestation across cultures, including those in North America, Europe, Korea, and Nigeria.

A similar convergence occurs in the aesthetics of tree form. Subjects in cross-cultural psychological tests prefer moderate-sized and sturdy trees with broad, layered canopies close to the ground. The species considered most attractive include acacias, which are dominant elements of healthy African savannas.

Tree aesthetics brings us to the question of the origin of the bio-

philic instincts. The human habitat preference is consistent with
the "savanna hypothesis," that humanity originated in the savannas and transitional forests of Africa. Almost the full evolutionary
history of the genus *Homo,* including *Homo sapiens* and its immediate ancestors, was spent in or near these habitats or others similar to them. If that amount of time, about two million years, were
to be compressed into a span of seventy years, humanity occupied
the ancestral environment for sixty-nine years and eight months,
whereupon some of the populations took up agriculture and
moved into villages to spend the last 120 days.

The savanna hypothesis extended to include behavior stipulates
that *Homo sapiens* is likely to be genetically specialized for the
ancestral environment so that today, even in the most sequestered
stone-and-glass cities, we still prefer it. Part of human nature is a
residue of bias in mental development that causes us to gravitate
back to savannas or their surrogates.

The savanna hypothesis of habitat preference may strike some
readers as evolutionism run amok. But is the idea really so strange?
Not at all: just a glance at the world of animal behavior suggests
otherwise. Every species that moves under its own power, from
protozoans to chimpanzees, instinctively seeks the habitat it must
occupy in order to survive and reproduce. The behavioral steps for
which it is genetically programmed are usually complex and
exactly executed. The study of habitat selection is an important
branch of ecology, and no species ever lets down the researcher
who chooses to examine this part of its life cycle. To take one of a
multitude of excellent examples, the African mosquito *Anopheles
gambiae* is a species specialized to feed on human blood. (As a
result it is a carrier of the malignant malarial parasite *Plasmodium
falciparum.*) Each female, in order to complete her life cycle, finds
her way from the stagnant pool of her birth and larval growth to a
nearby village. In the daytime she hides in crevices of the house.
At night she flies directly to one of the inhabitants, moving
upwind through a plume of the chemically distinctive odor of the

human body. She accomplishes all this with no experience and a brain the size of a grain of salt.

So it should be no great surprise that human beings, a biological species dependent on certain natural environments until very recently in its evolutionary history, should retain an aesthetic preference for savannas and transitional woodland among an array of natural and artificial environments laid before them. In general, what we call aesthetics may be just the pleasurable sensations we get from the particular stimuli to which our brains are inherently adapted.

To say that there is an instinct, or more accurately an array of instincts, that can be labeled biophilia is not to imply that the brain is hardwired. We do not ambulate like robots to the nearest lakeshore meadow. Instead, the brain is predisposed to acquire certain preferences as opposed to others. Psychologists who study mental development say that we are hereditarily *prepared* to learn certain behaviors and *counterprepared* to learn others. The vast majority of humans, to use a familiar example, are prepared to learn the lyrics of a song but counterprepared to learn calculus. We delight in the first and are fearful and begrudging of the second. Also, true to the pattern of instinct thus broadly defined, there are sensitive periods during childhood and early maturity in which learning and distaste are most easily picked up. In a manner also true to the conception, the timing varies among categories of behavior. Fluency in language comes earlier than fluency in mathematics.

The critical stages in the acquisition of biophilia have been worked out by psychologists during studies of childhood mental development. Under the age of six, children tend to be egocentric, self-serving, and domineering in their responses to animals and nature. They are also most prone to be uncaring or fearful of the natural world and of all but a few familiar animals. Between six and nine, children become interested in wild creatures for the first time, and aware that animals can suffer pain and distress.

From nine to twelve their knowledge and interest in the natural world rises sharply, and between thirteen and seventeen they readily acquire moral feeling toward animal welfare and species conservation.

A single study in the United States devoted to the subject suggests that a parallel sequence unfolds in the development of habitat preference. Children between the ages of eight and eleven, when given a choice of environmental photographs spread before them, favored savanna over hardwood forest, north-temperate conifer forest, rainforest, and desert. In contrast, older children preferred hardwood forest and savanna equally—in other words, habitats with which they had the most direct experience during their adolescence. Both of these environments were chosen over the remaining three. From this one set of data at least, the evidence supports the savanna hypothesis. In other words, children are evidently predisposed to favor the ancestral human habitat, but then increasingly favor the environment in which they have grown up.

Another sequence occurs in the way children explore the environment. At four they confine themselves to the immediate vicinity of their home and to small creatures readily found there, the "worms, chipmunks and pigeons" of neighboring yards and streets, as David Sobel expressed it in *Children's Special Places*. At eight to eleven they head for nearby woods, fields, ditches, and other unclaimed spots they can claim as their own. There they often build some kind of shelter such as a tree house, fort, or cave where they can read magazines, eat lunch, conspire with a friend or two, play games, and spy on the world. If natural wild environments are available, so much the better, but they are not essential. In urban East Harlem, children were observed building forts in culverts, alleyways, basements, abandoned warehouses, railroad right-of-ways, and hedges.

The secret places of childhood, whether a product of instinct or not, at the very least predispose us to acquire certain preferences and to undertake practices of later value in survival. The hideaways bond us with place, and they nourish our individuality and self-esteem. They enhance joy in the construction of habitation. If played out in natural environments, they also bring us close to the earth and nature in ways that can engender a lifelong love of both. Such was my own experience as a boy of eleven to thirteen, when I sought little Edens in the forests of Alabama and Florida. On one occasion I built a small hut of saplings in a remote off-trail spot. Unfortunately, I didn't notice until later that some of the saplings were poison oak, a virulent relative of poison ivy. That was the last of my secret-house constructions, but my love of the natural world nevertheless grew even stronger.

If biophilia is truly part of human nature, if it is truly an instinct, we should be able to find evidence of a positive effect of the natural world and other organisms on health. In fact, the annals of physiology and medicine contain abundant and diverse studies affirming just such a connection, at least when health is broadly defined, to use the words of the World Health Organization, as "a state of complete physical, mental and social well-being and not merely the absence of disease and infirmity." The following results of published studies are representative:

• A population of 120 volunteers were shown a stressful movie, followed by videotapes of either natural or urban settings. By their own subjective rating, they recovered from the feeling of stress more quickly while experiencing the natural settings. Their opinion was supported by four standard physiological measures of stress: heartbeat, systolic blood pressure, facial muscle tension, and electrical skin conductance. The results suggest, although don't prove, the involvement of the parasympathetic nerves, that part of the autonomic system whose activation induces a state of relaxed awareness. The same result was obtained in a different group of student volunteers stressed by a difficult mathematical examination, then shown videotapes that simulated automobile rides through natural as opposed to urban settings.

• Studies of response prior to surgery and dental work have consistently revealed a significant reduction of stress in the presence of plants and aquaria. Natural environments viewed through windows or merely displayed in wall-mounted pictures produce the same effect.

• Postsurgical patients recover more quickly, suffer fewer minor complications, and need smaller dosages of painkillers if given a window view of open terrain or waterscape.

• In one Swedish study covering fifteen years of records, clinically anxious psychiatric patients responded positively to wall pictures of natural environments, but negatively, occasionally even violently, to most other decorations (especially those containing abstract art).

• Comparable studies in prisons revealed that inmates provided window views of nearby farmlands and forests, as opposed to prison yards, reported fewer stress-related symptoms such as headaches and indigestion.

• In a different category, the popular notion that owning pets reduces stress-related problems has been well supported by research conducted independently in Australia, England, and the United States. In one Australian study, which factored out variation in exercise levels, diet, and social class, pet ownership accounted for a statistically significant reduction of cholesterol, triglycerides, and systolic blood pressure. In a parallel U.S. study, survivors of heart attacks (myocardial infarction) who owned dogs had a survival rate six times higher than those who did not. The same benefit was not, I am sorry to report, enjoyed by cat owners.

The implications of biophilia for preventive medicine are substantial. The biophilic instinct can be counted as one of humanity's fortunate irrationalities, like women's choice to have fewer children when economically secure, that deserve to be understood better and put to more practical use. It is a remarkable fact that while average life expectancy in the leading industrialized countries has risen to nearly eighty years, the contribution of preventive medicine, including the design of healthful and curative environments, has remained far below potential. Obesity, diabetes, melanoma, asthma, depression, hip fracture, and breast cancer have risen in frequency since 1980. Further, despite advances in scientific knowledge and public awareness, neither coronary atherosclerosis among young people nor acute myocardial infarction among the middle-aged and old has declined. All of these conditions can be delayed or even avoided by preventive measures that include, in most cases and to the point I wish to make, a reconnection to the natural world. As such they are cost-effective, amounting to no more than salvage of natural habitats, improvements in landscape design, and relocation of windows in public buildings.

Of course nature has a dark side too. The face it presents to humanity is not always friendly. Throughout most of human deep history there have been predators eager to snatch us for dinner; venomous snakes ready with a fatal, defensive strike to the ankle; spiders and insects that bite, sting, and infect; and microbes designed to reduce the human body to malodorous catabolic chemicals. The reverse side of nature's green-and-gold is the black-and-scarlet of disease and death.

The companion of biophilia is therefore biophobia. Like the responses of biophilia, those of biophobia are acquired by prepared learning. They vary in intensity among individuals according to heredity and experience. At one end of the scale are mild distaste and feelings of apprehension. At the other end are full-blown clinical phobias that fire the sympathetic nervous system and produce panic, nausea, and cold sweat. The innate biophobic intensities are most readily evoked by sources of peril that have existed in the natural world throughout humanity's evolutionary past. They include heights, close spaces, running water, snakes, wolves, rats and mice, bats, spiders, and blood. In contrast, prepared learning is unknown in response to knives, frayed electric wires, automobiles, and guns, which, although far deadlier today than the ancient perils of humankind, are too recent in evolutionary history to have been targeted by genetically prepared learning.

The defining properties of hereditary predisposition are multiple. One negative experience may be enough to trigger the response and permanently instill the fear. The critical stimulus can be unexpected and very simple—for example, the abrupt approach of an animal face, or the writhing of a serpent or serpent-like object nearby. The likelihood of imprinting is enhanced by already existing stressful conditions that surround the event. The learning can even be vicarious: just witnessing panic in another person or listening to a scary story can induce it in some people.

Those in whom the fear has been implanted respond almost instantly and subconsciously to subliminal images. When psychologists flashed pictures of snakes or spiders to subjects for only fifteen to thirty milliseconds, intervals too brief to be processed by the conscious mind, those previously conditioned adversely to these animals reacted with automatic muscle changes in the face within less than half a second. Although the response was easily detectable by the researchers, the subjects remained unaware that anything had happened at all.

Because aversive responses are so well defined, it has been possible to apply standard tests used in human genetics to determine whether variation in them among people has at least a partly genetic basis. The measure of choice is heritability, the standard used in studies of personality, obesity, neuroticism, and other traits that display complex variation in human populations. Heritability of a given trait is the percentage of variation among individuals in a population due to differences in genes among the individuals, as opposed to the percentage caused by differences in their environment. The heritability of innate aversion to snakes, spiders, insects, and bats respectively has been estimated to be about 30 percent, a common figure for human behavioral traits in general. The heritability of proneness to agoraphobia, an extreme aversion to crowds or open areas, is about 40 percent.

Another characteristic of prepared aversion is the existence of a sensitive period, which as in biophilic behavior is the interval in the normal life cycle when learning is easiest and the trait most apt to be established. In the case of ophidiophobia (snake), arachnophobia (spider), and other animal phobias, the onset occurs during childhood, with about 70 percent of cases occurring by ten years of age. In contrast, agoraphobia is an affliction of adolescents and young adults, triggered in 60 percent of the cases between fifteen and thirty years of age.

If elements of the natural world can sometimes paralyze modern humans by the evocation of ancient instincts, human instinct can and does wreak havoc on the natural world. Finding themselves surrounded by forests that once covered most of Earth's habitable land, Neolithic peoples set out ten thousand years ago to convert them into cropland, pasture, corrals, and scattered woodlots. What they could not chop down, they burned. Successive generations, their populations growing, continued the process until today only half the original cover is left. They needed the food, of course, but there is another way of looking at the relentless deforestation. People then as now instinctively wanted the ancestral habitat. So they proceeded to create savannas crafted to human needs. *Homo sapiens* did not evolve to be a forest dweller, like chimpanzees, gorillas, and other great apes. Rather, it became a specialist of open spaces. The aesthetically ideal environment of today's transformed world is the much-treasured pastoral landscape, for better or worse our ersatz savanna.

Where does attachment to that habitat leave wilderness? No question in environmental ethics cuts more deeply. Before agriculture and villages were invented, people lived in or very close to nature. They were part of it, and had no need for the concept of wilderness. Pastoral settlers drew a line between cultivated and virgin land. As they pushed back virgin land and built more complex societies with the aid of agricultural surpluses, they sharpened the distinction. Those in more advanced cultures imagined themselves to be above the untamed world around them. They were destined, they thought, to dwell among the gods. The word "wilderness" acquired the meaning expressed in its Old English progenitor *wil(d)dēornes:* wild, savage. To pastoral and urban sensibilities it was the impenetrable dark woods, the mountain fastness, the thornbush desert, the open sea, and any other part of the world that had not been and might never be tamed. It was the realm of beasts, savages, evil spirits, magic, and the menacing amorphous unknown.

The European conquest of the New World established the concept of wilderness as a frontier region waiting to be rolled back. The image was most clearly formed in the United States, whose early history is geographically defined as a westward march across an undeveloped and fertile continent.

Then came a tipping point. By the time the American frontier closed, around 1890, wilderness had become a scarce resource at risk of being eliminated altogether, and hence worth saving. American environmentalism was born, rising upon the new conservation ethic created by Henry David Thoreau, John Muir, and other nineteenth-century prophets. It spread slowly through the United States, Europe, and elsewhere. It argued that humanity would be foolish to wager its future on a wholly transformed planet. Wild lands in particular, the early environmentalists said, have a unique value for humankind. The warrior king of the movement was Theodore Roosevelt, who declared, "I hate a man who skins the land."

What is a wilderness today in our largely humanized world? What it has always been: a space that sustains itself, was here before humanity, and where, in the words of the Wilderness Act of 1964, "the Earth and its community of life are untrammeled by man and where man himself is a visitor who does not remain." The true great wildernesses of the world include the rainforests of the Amazon, the Congo, and New Guinea; the evergreen coniferous forests of northern North America and Eurasia; and Earth's ancient deserts, polar regions, and open seas.

A few contrarians like to claim that true wilderness is a thing of the past. They point out, correctly, that very few places on land have remained untrodden by human feet. Moreover, 5 percent of Earth's land surface is burned every year, and the plumes of nitrous oxide produced travel most of the way around the world. Greenhouse gases thicken, global temperatures rise, and glaciers and montane forests retreat up mountain peaks. With the exception of a few places in tropical Asia and Africa, terrestrial environments everywhere have lost most of their largest mammals, birds, and reptiles, destabilizing the populations of many other kinds of plants and animals. As the remnant wild areas shrink, they are invaded by more and more alien species, diminishing the native plants and animals yet more. The smaller the area of the natural reserves, the more we are forced to intervene to avoid the partial collapse of their ecosystems.

All true. But to claim that the surviving wildernesses are less than the name implies, and have in some sense become part of the human domain, is false. The argument is specious. It is like flattening the Himalayas to the level of the Ganges Delta by saying that all the planet's surface is but a geometer's plane. Walk from a pasture into a tropical rainforest, sail from a harbor marina to a coral reef, and you will see the difference. The glory of the primeval world is still there to protect and savor.

The exact perception of wilderness is a matter of scale. Even in disturbed environments, with most of their native plants and vertebrates long vanished, bacteria, protozoans, and miniature invertebrates still maintain the ancient substratum. The micro-wildernesses are more accessible than full-scale wildernesses. They are usually only minutes away, waiting to be visited by microscope instead of jetliner. A single tree in a city park, harboring thousands of species, is an island, complete with miniature mountains, valleys, lakes, and subterranean caverns. Scientists have only begun to explore these compacted worlds. Educators have made surprisingly little use of them in introducing the wonders of life to students. Microaesthetics based upon them is still an unexplored wilderness to the creative mind.

A strong case can be made for the creation of microreserves. A one-hectare patch of rainforest still clinging to a Honduran hillside, a roadstrip of native grasses in Iowa, and a muddy natural pond on the edge of a Florida golf course are to be valued and preserved even if the large native organisms that once lived in and around them have disappeared.

Still, while microreserves are infinitely better than nothing at all, they are no substitute for macro- and megareserves, where full-blown biotas with sizable animals continue to live. People can acquire an appreciation for savage carnivorous nematodes and shape-shifting rotifers in a drop of pond water, but they need life on the larger scale to which the human intellect and emotion most naturally respond. No one of my acquaintance, except a few microbiologists, would visit a town dump upon being told it harbors a dazzling variety of bacteria. But tourists and locals alike travel to the dumps of subarctic Canadian towns to watch scavenging polar bears.

To the multiple valorizations of wild environments can be added mystery. Without mystery life shrinks. The completely known is a numbing void to all active minds. Even a laboratory rat seeks the adventure of the maze.

So we are drawn to the natural world, aware that it contains structure and complexity and length of history as well, at orders of magnitude greater than anything yet conceived in human imagination. Mysteries solved within it merely uncover more mysteries beyond. For the naturalist every entrance into a wild environment rekindles an excitement that is childlike in spontaneity, often tinged with apprehension—in short, the way life ought to be lived, all the time.

I will offer one such personal remembrance out of hundreds forever fresh in my mind. It is the summer of 1965, in the Dry Tortugas, at the tip of the Florida Keys. I stand at the water's edge on Garden Key, with Fort Jefferson at my back, looking across a narrow channel to Bush Key, where the littoral scrub and mangrove swamp are alive with thousands of nesting sooty terns. I have a boat, and I will go there soon, but right now I have an inexplicable urge to swim across instead. The channel is about a hundred feet across, maybe less, and the tidal current from the Gulf of Mexico to Florida Bay is for the moment too slow to pose a risk. There will be no problem if I choose to swim, it seems. Then I look more closely at the moving water. How deep is the channel center? What might come up from below to meet me? A barracuda? I saw a five-footer circling the nearby dock pilings that morning. And what do I know about the local sharks? Hammerheads and bull sharks are common in deeper water, for sure, and have been known to attack humans. Great whites are occasionally seen. Shark attacks in this region are very rare, yet—would I be the dramatic exception? Now, reflecting as I hesitate, I feel an urge not just to cross, but to dive and explore the bottom of the channel. I want to know it inch by inch as well as I know the soil surface of the islands I have been studying, to see what else lives there and comes in sporadically from the Gulf.

The impulse to swim fades as quickly as it arose, but I make a resolution to come back someday and become an intimate of the channel and its inhabitants and to bond with this place on which I have randomly fixated, to make it part of my life. There is something crazy about the episode, but also something real, primal, and deeply satisfying.

At some time in our lives—for the naturalist always—we long for the gate to the paradisiacal world. It is the instinctive after-image that comes to us in daydreams, and a wellspring of hope. Its mysteries, if ignited in our minds and solved, grant more control over existence. If ignored, they leave an emotional void. How did such a strange quality of human nature come about? No one knows for sure, but evolutionary genetics tells us that even if just one person in a thousand survived because of a genetic predisposition to explore the unknown and persevere in daunting circumstances, then over many generations natural selection would have installed the predisposition in the whole human race to wonder and take the dare.

We need nature, and particularly its wilderness strongholds. It is the alien world that gave rise to our species, and the home to which we can safely return. It offers choices our spirit was designed to enjoy.

CHAPTER 7

———o———

THE SOLUTION

The human species is like the mythical giant Antaeus, who drew strength from contact with his mother, Gaea, the goddess Earth, and used it to challenge and defeat all comers. Hercules, learning his secret, lifted and held Antaeus above the ground until the giant weakened—then crushed him. Mortal humans are also handicapped by our separation from Earth, but our impairment is self-administered, and it has this added twist: our exertions also weaken Earth.

What humanity is inflicting on itself and Earth is, to use a modern metaphor, the result of a mistake in capital investment. Having appropriated the planet's natural resources, we chose to annuitize them with a short-term maturity reached by progressively increasing payouts. At the time it seemed a wise decision. To many it still does. The result is rising per-capita production and consumption, markets awash in consumer goods and grain, and a surplus of optimistic economists. But there is a problem: the key elements of natural capital, Earth's arable land, ground water, forests, marine fisheries, and petroleum, are ultimately finite, and not subject to proportionate capital growth. Moreover, they are being decapitalized by overharvesting and environmental destruction. With population and consumption continuing to grow, the per-capita resources left to be harvested are shrinking. The long-term prospects are not promising. Awakened at last to this approaching difficulty, we have begun a frantic search for substitutes.

Meanwhile, two collateral results of the annuitization of nature, as opposed to its stewardship, are settling in to beg our attention. The first is economic disparity: in relative terms the rich grow richer and the poor poorer. According to the United Nations Human Development Report 1999, the income differential between the fifth of the world's population in the wealthiest countries and the fifth in the poorest was 30 to 1 in 1960, 60 to 1 in 1990, and 74 to 1 in 1995. Wealthy people are also by and large profligate consumers, and as a result the income differential has this disturbing consequence: for the rest of the world to reach United States levels of consumption with existing technology would require four more planet Earths.

Europe is only slightly behind, while the Asian economic tigers appear to be pulling up at maximum possible speed. The income gap is the setting for resentment and fanaticism that causes even the strongest nations, led by the American colossus, to conduct their affairs with an uneasy conscience and a growing fear of heaven-bound suicide bombers.

The second collateral result, and the principal concern of the present work, is the accelerating extinction of natural ecosystems and species. The damage already done cannot be repaired within any period of time that has meaning for the human mind. The fossil record shows that new faunas and floras take millions of years to evolve to the richness of the prehuman world. The more the losses are allowed to accumulate, the more future generations will suffer for it, in some ways already felt and in others no doubt waiting to be painfully learned.

Why, our descendants will ask, by needlessly extinguishing the lives of other species, did we permanently impoverish our own? That hypothetical question is not the rhetoric of radical environmentalism. It expresses a growing concern among leaders in science, religion, business, and government, as well as the educated public.

What is the solution to biological impoverishment? The answer I will now pose is guardedly optimistic. In essence, it is that the problem is now well understood, we have a grip on its dimensions and magnitude, and a workable strategy has begun to take shape.

The new strategy to save the world's fauna and flora begins, as in all human affairs, with ethics. Moral reasoning is not a cultural artifact invented for convenience. It is and always has been the vital glue of society, the means by which transactions are made and honored to ensure survival. Every society is guided by ethical precepts, and every one of its members is expected to follow moral leadership and ethics-based tribal law. The propensity does not have to be beaten into us. Evidence exists instead of an instinct to behave ethically, or at least to insist on ethical behavior in others. Psychologists, for example, have discovered a hereditary tendency to detect cheaters and to respond to them with intense moral outrage. People by and large are natural geniuses at spotting deception in others, and equally brilliant in constructing deceptions of their own. We are daily soaked in self-righteous gossip. We pummel others with expostulation, and we hunger for sincerity in all our relationships. Even the tyrant is sterling in pose, invoking patriotism and economic necessity to justify his misdeeds. At the next level down, the convicted criminal is expected to show remorse, in the course of which he explains he was either insane at the time or redressing personal injustice.

And everyone has some kind of environmental ethic, even if it somehow makes a virtue of cutting the last ancient forests and damming the last wild rivers. Done, it is said, to grow the economy and save jobs. Done because we are running short of space and fuel. *Hey, listen, people come first!*—and most certainly before beach mice and louseworts. I recall vividly the conversation I had with a cab driver in Key West in 1968 when we touched on the

Everglades burning to the north. Too bad, he said. The Everglades are a wonderful place. But wilderness always gives way to civilization, doesn't it? That is progress and the way of the world, and we can't do much about it.

Everyone is also an avowed environmentalist. No one says flatly, "To hell with nature." On the other hand, no one says, "Let's give it all back to nature." Rather, when invoking the social contract by which we all live, the typical people-first ethicist thinks about the environment short-term and the typical environmental ethicist thinks about it long-term. Both are sincere and have something true and important to say. The people-first thinker says we need to take a little cut here and there; the environmentalist says nature is dying the death of a thousand cuts. So how do we combine the best of short-term and long-term goals? Perhaps, despite decades of bitter philosophical dispute, an optimum mix of the goals might result in a consensus more satisfactory than either side thought possible from total victory alone. Down deep, I believe, no one wants a total victory. The people-firster likes parks, and the environmentalist rides petroleum-powered vehicles to get there.

The first step is to turn away from claims of inherent moral superiority based on political ideology and religious dogma. The problems of the environment have become too complicated to be solved by piety and an unyielding clash of good intentions.

The next step is to disarm. The most destructive weapons to be stacked are the stereotypes, the total-war portraits crafted for public consumption by extremists on both sides. I know them very well from years of experience on the boards of conservation organizations, as a participant in policy conferences, and during service on government advisory committees. To tell the truth, I am a little battle-fatigued. The stereotypes cannot be simply dismissed, since they are so often voiced and contain elements of real substance, like rocks in snowballs. But they can be understood clearly and sidestepped in the search for common ground. Let me illustrate a stereotype skirmish with imaginary opponents engaging in typical denunciations.

THE PEOPLE-FIRST CRITIC STEREOTYPES
THE ENVIRONMENTALISTS

Environmentalists or conservationists is what they usually call themselves. Depending on how angry we are, we call them greens, enviros, environmental extremists, or environmental wackos. Mark my word, conservation pushed by these people always goes too far, because it is an instrument for gaining political power. The wackos have a broad and mostly hidden agenda that always comes from the left, usually far left. How to get power? is what they're thinking. Their aim is to expand government, especially the federal government. They want environmental laws and regulatory surveillance to create government-supported jobs for their kind of bureaucrats, lawyers, and consultants. The New Class, these professionals have been called. What's at stake as they busy themselves are your tax dollars and mine, and ultimately our freedom too. Relax your guard when these people are in power and your property rights go down the tube. Some Bennington College student with a summer job will find an endangered red spider on your property, and before you know what happened the Endangered Species Act will be used to shut you down. Can't sell to a developer, can't even harvest your woodlot. Business investors can't get at the oil and gas on federal lands this country badly needs. Mind you, I'm all for the environment, and I agree that species extinction is a bad thing, but conservation should be kept in perspective. It is best put in private hands. Property owners know what's good for their own land. They care about the plants and animals living there. Let them work out conservation. They are the real grass roots in this country. Let them be the stewards and handle conservation. A strong, growing free-market economy, not creeping socialism, is what's best for America—and it's best for the environment too.

THE ENVIRONMENTALIST STEREOTYPES
THE PEOPLE-FIRST CRITICS

"Critics" of the environmental movement? That may be what they call themselves, but we know them more accurately as anti-environmentalists and brown lashers or, more locally out west, wise users (their own term, not intended to be ironic) and sagebrush rebels. In claiming concern of any kind for the natural environment, these people are the worst bunch of hypocrites you'll ever not want to find. What they are really after, especially the corporate heads and big-time landowners, is unrestrained capitalism with land development über alles. They keep their right-wing political agenda mostly hidden when downgrading climate change and species extinction, but for them economic growth is always the ultimate, and maybe the only, good. Their idea of conservation is stocking trout streams and planting trees around golf courses. Their conception of the public trust is a strong military establishment and subsidies for loggers and ranchers. The anti-environmentalists would be laughed out of court if they weren't tied so closely to the corporate power structure. And notice how rarely international policy makers pay attention to the environment. At the big conferences of the World Trade Organization and other such gatherings of the rich and powerful, conservation almost never gets so much as a hearing. The only recourse we have is to protest at their meetings. We hope to attract the attention of the media and at least get our unelected rulers to look out the window. In America the right-wingers have made the word "conservative" a mockery. What exactly are they trying to conserve? Their own selfish interests, for sure, not the natural environment.

There are partisans on both sides who actually state their case in this manner, either in pieces or in entirety. And the accusations sting, because so many people on either side believe them. The suspicion and anger they express paralyze further discussion. Worse, in an era when journalism feeds on controversy, its widely used gladiatorial approach divides people and pushes them away from the center toward opposite extremes.

It is a contest that will not be settled by partisan victory. The truth is that everyone wants a highly productive economy and lots of well-paying jobs. People almost all agree that private property is a sacred right. On the other hand, everyone treasures a clean environment. In the United States at least, the preservation of nature has almost the status of a sacred trust. In a 1996 survey conducted by Belden & Russonello for the U.S. Consultative Group on Biological Diversity, 79 percent rated a healthy and pleasant environment of the greatest importance, giving it a 10 on a scale of 1 to 10. Seventy-one percent agreed at the same high level with the statement "Nature is God's creation and humans should respect God's work." Only when these two obvious and admirable goals, prosperity and saving the creation, are cast in opposition does the issue become confused. And when the apparent conflict is in addition reinforced by opposing political ideologies, as it frequently is, the problem becomes intractable.

The ethical solution is to diagnose and disconnect extraneous political ideology, then shed it in order to move toward the common ground where economic progress and conservation are treated as one and the same goal.

The guiding principles of a united environmental movement must be, and eventually will be, chiefly long-term. If two hundred years of history of environmentalism have taught us anything, it is that a change of heart occurs when people look beyond themselves to others, and then to the rest of life. It is strengthened when they also expand their view of landscape, from parish to nation and beyond, and their sweep of time from their own life spans to multiple generations and finally to the extended future history of humankind.

The precepts of the people-firsters are foundationally just as ethical as those of the traditional environmentalists, but their arguments are more about method and short-term results. Further, their values are not, as often assumed, merely a reflection of capitalist philosophy. Corporate CEOs are people too, with families and the same desire for a healthy, biodiverse world. Many are leaders in the environmental movement. It is time to recognize that their commitment is vital to success. The world economy is now propelled by venture capital and technical innovation; it cannot be returned to a pastoral civilization. Nor will socialism return in a second attempt to rescue us, at least in any form resembling the Soviet model. Quite the contrary, its demise was a good thing all around for nature. In most places the socialist experiment was tried, its record was even worse than in capitalist countries. Totalitarianism, left or right, is a devil's bargain: slavery purchased at the price of a ruined environment.

The juggernaut of technology-based capitalism will not be stopped. Its momentum is reinforced by the billions of poor people in developing countries anxious to participate in order to share the material wealth of the industrialized nations. But its direction can be changed by mandate of a generally shared long-term environmental ethic. The choice is clear: the juggernaut will very soon either chew up what remains of the living world, or it will be redirected to save it.

Science and technology are themselves reason for optimism. They are growing exponentially—in the case of computer capacity, superexponentially, the doubling time having dropped to one year. The consequences are not predictable, but one is almost certainly to be an improvement of human self-understanding. Within several decades, many neuroscientists believe, we will have a much firmer grasp of the biological sources of mind and behavior. That in turn will provide the basis for a more solid social science, and a better capacity to anticipate and step away from political and economic disasters.

Also emerging swiftly is a sophisticated picture of changes in global environment and available resources. Concrete measures such as the ecological footprint and the Living Planet Index form the groundwork for wiser economic planning. Science and technology also promise the means for raising per-capita food production while decreasing materials and energy consumption, both of which are preconditions for successful long-term conservation and a sustainable economy.

All this information is coming on-line worldwide. It will allow people everywhere to see the planet as the astronauts see it, a little sphere with a razor-thin coat of life too fragile to bear careless tampering. A growing cadre of leaders in business, government, and religion now think in this foresighted manner. They understand that humanity is in a bottleneck of overpopulation and wasteful consumption. They agree, at least in principle, that we will have to maneuver carefully in order to pass through the bottleneck safely.

To lift a stabilized world population to a decent quality of life while salvaging and restoring the natural environment is a noble and attainable goal. This brings me to another source of cautious optimism, the growing prominence of the environment in religious thought. The trend is important not only for its moral content, but for the conservatism and authenticity of its nature. Religious leaders are by necessity very careful in the values they choose to promote. The sacred texts from which they draw authority tolerate few amendments. In modern times, as knowledge of the material world and the human predicament has soared, the leaders have followed rather than led the evolution of ethics. First into the new terrain venture saints and radical theologians. They are followed by growing numbers of the faithful and then, warily, by the bishops, patriarchs, and imams.

For the Abrahamic religions, Judaism, Christianity, and Islam, the environmental ethic is compatible with belief in the holiness of the Earth and the perception of nature as God's handiwork. In the thirteenth century Saint Francis of Assisi prayed for the welfare of God's creatures, his avowed "brothers and sisters," and extolled the "beautiful relationship" of humankind and nature. In Genesis 1:28, God instructs Adam and Eve to "fill the earth, and subdue it, and rule over the fishes of the sea, and the birds of the air, and all living creatures that move upon the earth." Once in history the passage was construed to validate the conversion of nature to exclusively human needs. Now it is more commonly interpreted to mean the stewardship of nature. Thus Pope John Paul II has affirmed that "the ecological crisis is a moral issue." And Patriarch Bartholomew I, spiritual leader of the world's 250 million Orthodox Christians, has declared, in the clarion tones of an Old Testament prophet, that "for humans to cause species to become extinct and to destroy the biological diversity of God's creation, for humans to degrade the integrity of the earth by caus-ing changes in its climate, by stripping the earth of its natural forests, or destroying its wetlands, for humans to contaminate the earth's waters, its land, its air and its life with poisonous sub-stances, these are sins."

Some Protestant denominations are active in conservation. Among them are evangelical sects prone to literal interpretations of the Bible. In 1988 the Reverend Stan L. LeQuire, director of the Evangelical Environmental Network, stated the issue incisively: "We evangelicals are recognizing more and more that environmen-tal issues are not Republican or Democratic, that they really come from the most wonderful teachings that we have in Scripture, which commend us to honor God by caring for creation." His network, organized into "Noah Congregations," proved its met-tle: it contributed $1 million to the successful campaign against congressional efforts to weaken the Endangered Species Act.

In the evangelical culture God can still strike the wicked, even if only through the dire consequences of their own actions. Listen to the voice of Janisse Ray, a young poet from southern Georgia, who in her 1999 memoir *Ecology of a Cracker Childhood* decries the destruction of the region's longleaf pine forests. Her warning perfectly captures the cadence of an evangelical sermon:

If you clear a forest, you'd better pray continuously. While you're pushing a road through and rigging the cables and mov-ing between trees on the dozer, you'd better be talking to God. While you're cruising timber and marking trees with a blue slash, be praying; and pray while you're peddling the chips and

logs and writing Friday's checks and paying the diesel bill—
even if it's under your breath, a rustling at the lips. If you're
manning the saw head or the scissors, snipping the trees off
at the ground, going from one to another, approaching them
brusquely and laying them down, I'd say, pray extra hard; and
pray hard when you're hauling them away.

God doesn't like a clearcut. It makes his heart turn cold,
makes him wince and wonder what went wrong with his creation, and sets him to thinking about what spoils the child.

A scattering of Roman Catholic dioceses and Jewish syna-
gogues have also joined in environmental activism. The Religious
Campaign for Forest Conservation, an interfaith group founded
in 2000, aims at uniting Jewish and Christian efforts. Its members
share a conviction that activities destroying the natural environment "foster injustice and gross economic inequities. Most griev-
ously we declare them spiritually bankrupt because they deny God
and foster the degradation of human society."

On one memorable occasion, in the fall of 1986, I was invited
by the Committee on Human Values of the U.S. Roman Catholic
Bishops to discuss the relation of science to religion. I was joined
at their two-day retreat, held near Detroit, by three other scientists
and a group of Catholic lay theologians. As one professor of theol-
ogy expressed it, "Science went out the door with Aquinas and we
never invited her back." The times were changing. At the end of
our varied and frank discussions the bishops drew up a list of pri-
orities for postconference study. Second from the top was environment and conservation.

On a later occasion, also symptomatic of the trend toward
moral consensus, I was invited by my friend Bruce Babbitt, then
secretary of the interior under President Clinton, to join him,
another scientist, and several religious leaders for a discussion of
the role of our respective callings in promoting conservation. The
atmosphere was wholly congenial, even faintly conspiratorial. As
we closed, Babbitt remarked that if the two most powerful forces

in America, religion and science, could be united on the issue, the country's environmental problems would be quickly solved.

Such a collaboration is feasible. I like to think that the environmental values of secular and religious alike arise from the same innate attraction to nature. They express the same compassion for animals, aesthetic response to free-living flowers and birds, and wonder at the mysteries of wild environments. Of course, secular and religious thinkers differ as ever in their explanations of where these feelings originate. They argue about who, or Who, judges the stewardship that a common ethic commands. But these epistemological distinctions, so important in other spheres of public life, can be safely put aside in the case of the environment. Polls show that, in the United States at least, people of all socioeconomic groups and religious beliefs become conservationists when well informed, and primarily for moral reasons. Even despoilers pay tribute to virtue. They assure us that under certain conditions the logging of old trees reduces fire damage and helps wildlife. They protest that maybe global warming won't be so bad after all. They allow that protecting pandas and gorillas and eagles is a good idea.

The convergence in opinion is strong enough that the problem is no longer the reasons for conservation but the best method to achieve it. That challenge, while enormous, can be met. During the past two decades, scientists and conservation professionals have put together a strategy aimed at the protection of most of the remaining ecosystems and species. Its key elements are the following:

- Salvage immediately the world's hotspots, those habitats that are both at the greatest risk and shelter the largest concentrations of species found nowhere else. Among the most valuable hotspots on the land, for example, are the surviving remnants of rainforest in Hawaii, the West Indies, Ecuador, Atlantic Brazil, West Africa, Madagascar, the Philippines, Indo-Burma, and India, as well as the Mediterranean-climate scrublands of South Africa, southwestern Australia, and southern California.

Twenty-five of these special ecosystems cover only 1.4 percent of Earth's land surface, about the same as Texas and Alaska combined. Yet they are the last remaining homes of an impressive 43.8 percent of all known species of vascular plants and 35.6 percent of the known mammals, birds, reptiles, and amphibians. The twenty-five hotspots have already been reduced 88 percent in area by clearing and development; some could be wiped out entirely within several decades by continued intrusion.

• Keep intact the five remaining frontier forests, which are the last true wildernesses on the land and home to an additional large fraction of Earth's biological diversity. They are the rainforests of the combined Amazon Basin and the Guianas; the Congo block of Central Africa; New Guinea; the temperate conifer forests of Canada and Alaska combined; and the temperate conifer forests of Russia, Finland, and Scandinavia combined.

• Cease all logging of old-growth forests everywhere. For every bit of this habitat lost or degraded, Earth pays a price in biodiversity. The cost is especially steep in tropical forests, and it is potentially catastrophic in the forested hotspots. At the same time, let secondary native forests recover. The time has come—rich opportunity shines forth—for the timber-extraction industry to shift to tree farming on already converted land. The cultivation of lumber and pulp should be conducted like the agribusiness it is, using high-quality, fast-growing species and strains for higher productivity and profit. To that end, it would be valuable to forge an international agreement, similar to the Montreal and Kyoto Protocols, that prohibits further destruction of old-growth forests and thereby provides the timber-extraction economy with a level playing field.

• Everywhere, not just in the hotspots and wildernesses, concentrate on the lakes and river systems, which are the most threatened ecosystems of all. Those in tropical and warm-temperate regions in particular possess the highest ratio of endangered species to area of any kind of habitat.

• Define precisely the marine hotspots of the world, and assign them the same action priority as for those on the land. Foremost are the coral reefs, which in their extremely high biological diversity rank as the rainforests of the sea. More than half around the world—including, for example, those of the Maldives and parts of the Caribbean and Philippines—have been savaged variously by overharvesting and rising temperatures, and are in critical condition.

• In order to render the conservation effort exact and cost-effective, complete the mapping of the world's biological diversity. Scientists have estimated that 10 percent or more of flowering plants, a majority of animals, and a huge majority of microorganisms remain undiscovered and unnamed, hence of unknown conservation status. As the map is filled in, it will evolve into a biological encyclopedia of value not only in conservation practice but also in science, industry, agriculture, and medicine. The expanded global biodiversity map will be the instrument that unites biology.

• Using recent advances in mapping the planet's terrestrial, fresh-water, and marine ecosystems, ensure that the full range of the world's ecosystems are included in a global conservation strategy. The scope of conservation must embrace not only the habitats, such as tropical forests and coral reefs, that harbor the richest assemblages of species, but also the deserts and arctic tundras whose beautiful and austere inhabitants are no less unique expressions of life.

• Make conservation profitable. Find ways to raise the income of those who live in and near the reserves. Give them a proprietary interest in the natural environment and engage them professionally in its protection. Help raise the productivity of land already converted to cropland and cattle ranches nearby, while tightening security around the reserves. Generate sources of revenue in the reserves themselves. Demonstrate to the governments, especially of developing countries, that ecotourism, bioprospecting, and (eventually) carbon credit trades of wild land can yield more income than logging and agriculture of the same land cleared and planted.

• Use biodiversity more effectively to benefit the world economy as a whole. Broaden field research and laboratory biotechnology to develop new crops, livestock, cultivated food fish, farmed timber, pharmaceuticals, and bioremedial bacteria. Where genetically engineered crop strains prove nutritionally and environmentally safe upon careful research and regulation, as I outlined in chapter 5, they should be employed. In addition to feeding the hungry, they can help take the pressure off the wildlands and the biodiversity they contain.

• Initiate restoration projects to increase the share of Earth allotted to nature. Today about 10 percent of the land surface is protected on paper. Even if rigorously conserved, this amount is not enough to save more than a modest fraction of wild species. Large numbers of plant and animal species are left with populations too small to persist. Every bit of space that can be added will pass more species through the bottleneck of over-population and development for the benefit of future generations. Eventually, and the sooner the better, a higher goal can and should be set. At the risk of being called an extremist, which on this topic I freely admit I am, let me suggest 50 percent. Half the world for humanity, half for the rest of life, to create a planet both self-sustaining and pleasant.

• Increase the capacity of zoos and botanical gardens to breed endangered species. Most are already working to fill that role. Prepare to clone species when all other preservation methods fail. Enlarge the existing seed and spore banks and create reserves of frozen embryos and tissue. But keep in mind that these methods are expensive and at best supplementary. Moreover, they are not feasible for the vast majority of species, especially the countless bacteria, archaeans, protistans, fungi, and insects and other invertebrates that make up the functioning base of the biosphere. And even if somehow, with enormous effort, all these species too could be stored artificially, it would

be virtually impossible to reassemble them later into sustainably free-living ecosystems. The only secure way to save species, as well as the cheapest (and on the evidence the only sane way), is to preserve the natural ecosystems they now compose.
• Support population planning. Help guide humanity everywhere to a smaller biomass, a lighter footstep, and a more secure and enjoyable future with biodiversity flourishing around it.

Earth is still productive enough and human ingenuity creative enough not only to feed the world now but also to raise the standard of living of the population projected to at least the middle of the twenty-first century. The great majority of ecosystems and species still surviving can also be protected. Of the two objectives, humanitarian and environmental, the latter is by far the cheaper, and the best bargain humanity has ever been offered. For global conservation, only one-thousandth of the current annual world domestic product, or $30 billion out of approximately $30 trillion, would accomplish most of the task. One key element, the protection and management of the world's existing natural reserves, could be financed by a one-cent-per-cup tax on coffee.

Progress toward global conservation, acceptance of the bargain, will pick up or falter depending on cooperation among the three secular stanchions of civilized existence: government, the private sector, and science and technology.

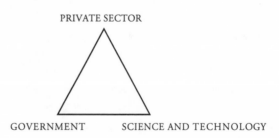

Governments devise laws and regulatory practices that, if ethically based, give long-term benefit to the governed. They treat the environment as a public trust. They are moreover party to treaties that protect the planetary environment as a whole, such as the 1982 United Nations Convention on the Law of the Sea, the 1987 Montreal Protocol on Substances That Deplete the Ozone, and the Convention on Biological Diversity of the 1992 Rio Earth Summit. The private sector, working within the public-trust constraints defined by government policy, is the engine of society. A strong economy improves the material quality of life, allowing the populace to look and plan ahead in all venues important to them, including the environment. In the process it buoys science and technology, the means to improve knowledge of the material world, control our own lives, and secure the prerequisites of personal fulfillment.

The interlocking of the three key agents is vital to global conservation. The trends in their evolution are encouraging. A legion of private and government-supported institutions is pursuing environmental initiatives unimagined twenty years ago. Public support of conservation, once tenuous around the world, has begun to quicken. Mexico, Ecuador, Brazil, Papua New Guinea, and Madagascar are among the developing countries with national programs directed at the preservation of natural habitats that most need them.

The spearhead of the global conservation movement consists of the nongovernmental organizations (NGOs). They range in size from the relatively giant Conservation International, Wildlife Conservation Society, World Wide Fund for Nature, World Wildlife Fund–U.S., and The Nature Conservancy to much smaller, specialized groups, of which the following is a representative sample: the Seacology Foundation (island environments and culture), Ecotrust (temperate rainforests of North America), Xerces Society (insects and other invertebrates), Bat Conservation International, and the Balikpapan Orangutan Society–USA. In 1956 there were, according to the Union of International Organizations, 985 NGOs devoted to humanitarian or environmental causes or both. By 1996 the number had increased to more than

20,000. Their membership and coordination has expanded throughout that period, a trend now enhanced by Internet advertising and communication. In the late 1990s there was one paid membership in some environmental organization or other for every twenty Americans, and more than one such membership for every citizen of Denmark. The governing boards and advisory committees bring together scientists, corporate executives, private investors, media stars, and other private citizens devoted to the cause.

The swift ascendancy of the NGOs reflects the perception within the global conservation movement that the extinction crisis has turned critical. As in a desperate battle with retreat not an option, the motivation exists to try new strategies. Those able to invent them take the lead. For the most part governments are hesitant, even timid. They are too distracted by the necessities of military defense, political intrigue, and the economy to deal effectively with the death of nature. Ordinary people care about the environment, but what they have in mind is chiefly pollution and climate change. Although they support conservation at home, those in the wealthier industrial countries have limited concern for biodiversity in the developing world, where the hemorrhaging is greatest. The idea of spending tax dollars on national parks in Peru and Vietnam remains too alien an idea for most to take seriously.

Into this gap the international nongovernmental organizations have moved, drawing on both their own and governmental resources. Funding is on the rise. And while grassroots support of the NGOs by membership and subscription is vital, a disproportionate share has been picked up by a small number of the wealthiest citizens and the corporations whose investments they control. The two hundred richest corporations of the world are virtually an empire unto themselves, with resources equal to the combined wealth of the poorest 80 percent of the world's population. Their leaders and principal shareholders, who occupy the commanding heights of economic and political power, are likely also to have the education and awareness to understand the environmental and humanitarian issues of global conservation. Attracted by inspired leadership within the NGOs, an increasing number have responded with their wealth and time.

One of the flagship NGOs that has benefited from the generosity of both regular members and major private donors has been the World Wildlife Fund.* As a member of its board of directors for ten years, from 1984 to 1994, I had the privilege of witnessing its phenomenal growth in funding and influence. During that period membership jumped from about 100,000 to more than a million, then leveled off as competitors alongside grew enough to saturate the support market. It was also a time of equally rapid evolution in the self-image and operational methods of the World Wildlife Fund and other large conservation organizations. In the early 1980s WWF was focused on what is known in the trade as the charismatic megafauna—pandas, rhinos, great cats, bears, eagles, and other instantly recognizable large animals, together with the habitats needed to sustain them. The rationale was primarily aesthetic, similar to that for preserving historical buildings and scenic vistas. Soon, however, the vision of WWF began a radical change. Its orientation turned from top-down to bottom-up. Now, entire ecosystems harboring the charismatic animals, which often contain hundreds of other endangered but less famous species, became the focus. Lesser-known hotspots were added, even those lacking familiar large animals—just so long as they harbored a goodly number of endangered species. The World Wildlife Fund never forgot the panda and the tiger and other symbolically potent clients. But its crusade steadily broadened until it included all of the world's threatened biodiversity.

---

*World Wildlife Fund–U.S. received its name as an affiliate of World Wildlife Fund–International, which is headquartered in Gland, Switzerland. When the latter conglomorate changed its name to the World Wide Fund for Nature, the U.S. organization retained the title World Wildlife Fund, without the identifying national suffix. The result is a taxonomic confusion: both organizations use the acronym WWF (unhappily also shared with the World Wrestling Federation), and both have kept the familiar stylized giant panda for their logo. The World Wildlife Fund remains a national affiliate of the World Wide Fund for Nature, of which it is the largest branch. The national units of the World Wide Fund for Nature together employ more than three thousand people and have combined revenues in excess of $300 million.

A second change by WWF was to form partnerships with the people who live in and around the targeted ecosystems. In addition to simple humanitarianism—most of the residents are impoverished—biodiversity conservation was well served, given the commonsense generality that no one can be expected to leave a reserve inviolate if it is his source of food and fuel. A patch of forest fenced off and patrolled is a cruel insult to hungry people shut out, and unworkable in the long run. That point was made bluntly by the Cameroonian journalist François Bikoro in response to recent attempts by the World Bank and the World Wide Fund for Nature to halt logging in the Central African rainforests: "You destroyed your environment and got developed. Now you want us to stop doing the same! What do we get out of it? You have your televisions and your cars but no trees. People want to know what they gain by conserving the forest." To which Claude Martin, director-general of the World Wide Fund for Nature, responded with the larger picture: When the great forest is gone, possibly by 2020 at the present rate of cutting, there will be no more jobs. Land cut over in the region is mostly abandoned, and poverty within it is greater than before. But local people with families to feed do not see the larger picture, and their needs cannot be met by a purely preservationist policy.

So a new goal was added by WWF and other organizations: develop a strategic mix of support and development to turn the reserve into an economic asset. Make the local people partners; give them an incentive to be stewards and guards of the reserve. Train them to be guides and resident wildlife experts. Persuade their government to see the reserve as a national treasure and source of income.

In planning global strategies it also became obvious to WWF and other NGOs that in order to save entire ecosystems it is also necessary to possess a great deal of scientific knowledge about them. Which habitats are both richest in biodiversity and most endangered? What is the minimal acreage to sustain them, the effects of physical disturbance, the impact of invasive species? And

then there are the people around the reserves, the future conservation partners. What are their political and economic circumstances, their customs, their environmental beliefs, their special
requirements? WWF responded by building its own research program, staffed by experts who work with the organization's regional
administrations in choosing and managing projects. For its part,
The Nature Conservancy (TNC) sponsored the Natural Heritage
Program, with the aim of registering all plant and animal species
of the United States and subsequently, as the independent Asso-
ciation for Biodiversity Information, all the threatened species of
the Western Hemisphere. Conservation International introduced
the Rapid Assessment Program to speed the mapping of poorly
known hotspot and wilderness areas, followed by the Center for
Applied Biodiversity Sciences (CABS) to support in-house research from systematics and ecology to economics and anthropology. CABS further instituted an unprecedented engagement with
the scholarly community, not only freely exchanging data but
channeling up to half of its program funding to other organizations. The financing of such alliances has raised the effectiveness
and credibility of conservation science.

All the major conservation organizations have evolved in an
approximately parallel manner and enjoyed robust growth. By
1999 the memberships of the six largest U.S.-based environmental
groups were as follows:

World Wildlife Fund	1,200,000
The Nature Conservancy	1,021,000
National Wildlife Federation	835,000
Sierra Club	392,000
National Parks Conservation Association	390,000
National Audubon Society	385,000

These organizations, together with Conservation International,
which relies on a wealthier donor base and has fewer members,
were operating on annual budgets at the $50 million to $100 mil-

lion level. In March 2000 The Nature Conservancy upped the ante by initiating a three-year capital campaign to raise and invest $1 billion in reserve acquisition. The organization's goal is to conserve two hundred key natural areas in the United States and abroad and to improve reserves already in its care. TNC's competence to deal at this level is supported by its track record: in 1998–9 it either acquired as gifts or purchased a total of 900,000 acres of environmentally valuable ecological land in the United States, bringing the total for its entire forty-eight-year history to 11.5 million acres, an area the size of Switzerland.

In 2001 Conservation International received a gift of $52.8 million from the Gordon E. and Betty I. Moore Foundation to expand research and reserve capacity in tropical wildernesses and hotspots. The World Wildlife Fund has also moved toward a higher level of funding, with larger stakes on the table. In 1997 President Fernando Henrique Cardoso of Brazil asked WWF to collaborate in planning and funding the management of a new system of eighty parks drawn from already existing public lands in the Amazon, spreading over 100 million acres, or 10 percent of the region, an area greater than the state of California. The price tag to sustain the park system in perpetuity is $270 million. Logging and mining are to be forbidden, and hunting and fishing permitted only for aboriginal inhabitants. The project was launched in 2001 and is to be built incrementally over a period of ten years, with funding coming primarily from multiple international aid and lending organizations.

In the 1970s, when I joined a small group of scientists active in the global conservation movement, including, among others, Paul Ehrlich, Thomas Lovejoy, Norman Myers, Peter Raven, and George Schaller, the role played by the nongovernmental groups we advised was basically that of evangelists and beggars. These organizations pointed to the plight of the world's declining fauna and flora. They listed and characterized many of the endangered species, a task accomplished with special authority by the Red List books of the International Union for the Conservation of Nature

and Natural Resources, or IUCN, also known as the World Conservation Union. The early global conservation groups conducted
campaigns as best they could on small budgets, in bits and pieces
here and there, and most successfully when pandas and tigers
and other charismatic species were the centerpieces. They traveled
upcurrent against skepticism and indifference to their campaign
on behalf of the natural world. I found it a galling experience to
have to plead like a defense attorney in court on behalf of biodiversity, justifying its existence, asking that it be spared. I still feel
that way, and especially when forced to do so on occasion in my
own country.

To the global conservationists in the early era the destruction of
natural environments and the species in them seemed almost
unstoppable. The greatest ongoing damage was and remains the
destruction of tropical forests, where most kinds of plants and animals on Earth live. It has always been clear that the struggle to save
biological diversity will be won or lost in the forests. By the 1970s,
when we first took a careful look around, half the forests were
already gone, and the annual rate of worldwide clearing was 1 to 2
percent of cover still intact in each succeeding year. By 2000, it
appeared to have subsided somewhat, to an estimated 33.8 million
acres per annum, or slightly under 1 percent, of the 3.5 billion
acres remaining. Optimism must be tempered, however, because
some of the decrease is due to the growing scarcity of accessible logging tracts. Some rainforests, including those in parts of
Indonesia and West and Central Africa, still suffer accelerating
losses. Also under heavy pressure are the rich hardwood and coniferous woodlands of western China and the southern face of the
Himalayas. Nepal, once gloriously clothed, has been largely
denuded.

By the 1990s the major global NGOs had grown strong enough
to initiate direct action on their own toward the salvaging of
forests and other threatened natural environments. Taking their
place at the table of business and government, they forged partnerships with corporations, heads of government, and interna-

tional lending and aid organizations to advance their large-scale agenda.

The NGOs also became more inventive. They recognized that the amount of protected land and shallow marine areas is far short of that needed to save all, or even most, of the world's biodiversity. At the same time they discovered that in many parts of the world, and especially in tropical forested countries where most of biodiversity exists, reserves can be enlarged or created at relatively low costs. The NGOs seized upon this circumstance to initiate arrangements both environmentally and economically attractive to the host countries. One of the first innovations, introduced in the 1980s, was the debt-for-nature swaps. The idea is remarkably simple: raise funds to purchase a portion of the country's commercial debt at a discount, or persuade creditor banks to donate some of it. Then exchange the debts in local currency on bonds at favorable rates. Such transfers are widely feasible because so many developing countries are close to default. Finally, the enlarged local equity is used to promote conservation, variously by purchase of land for reserves, environmental education, and improvement of management of existing reserves. By the early 1990s twenty such agreements, totaling $110 million, had been set up in Bolivia, Costa Rica, the Dominican Republic, Ecuador, Mexico, Madagascar, Zambia, the Philippines, and Poland.

In the late 1990s and start of the millennium a cluster of new initiatives was introduced, beginning a true revolution in global conservation. One of the most important is the conservation concession, which allows rapid set-asides of large blocks of tropical forest. It is, in the words of the ecological economist Richard Rice, "warp-speed conservation." A concession is a lease on a parcel of land granted by a government for a specific purpose. In the past, by far the largest such contracts made by developing countries have gone to logging companies, usually foreign-based, whose sole purpose is to clearcut and harvest the timber. So entrenched and seemingly profitable was this practice, so powerful the timber-extraction industry, that the destruction of forests everywhere seemed inevitable. Not so, it turns out. The profit margin of the logging companies is very thin in most tropical countries, forcing them to make offers of only a few dollars per acre. They can be outbid by determined conservation NGOs.

The first conservation concession was obtained in 2000 by Conservation International (CI) from Guyana, a small former British colony on the north coast of South America. Guyana's chief asset, and a source of national pride, is its interior wilderness of mostly pristine rainforest. For an application fee of $20,000 and fifteen cents an acre annually, Conservation International leased a 200,000-acre tract in the remote southeastern corner of the country. CI put up additional funds for management of the property as a nature reserve. The period is for three years, during which both parties will negotiate the rate for a subsequent twenty-five-year period. Amerindians in the area will be allowed to continue hunting, fishing, and conducting small-scale agriculture at the level they have practiced for thousands of years.

Guyana draws multiple benefits from the arrangement. It makes at least as much money as it would from a timber lease. It does so while holding on to its beautiful natural environment. And it has time to find noninvasive ways to produce still more income, including tourism, prospecting for useful plant products, and sustainable harvesting of plant material for medicinal use. With intact forests it may also someday enjoy the opportunity to sell carbon sequestration credits, an arrangement set forth by the Kyoto Climate Protocol as one device to reduce carbon dioxide and other greenhouse gases in the planet's atmosphere. In the arrangement, poor countries can receive money for merely saving their forests.

Encouraged by this initial success, Conservation International has begun (as I write, in early 2001) similar negotiations with Bolivia, Brazil, Peru, Cambodia, Indonesia, and Madagascar. All have agreed in principle to arrangements similar to the Guyana model.

Other initiatives are under way. In some cases large-scale conservation has been achieved by the outright purchase of logging rights. In 1998 The Nature Conservancy doubled the size of Bolivia's Noel Kempff Mercado National Park by acquiring the rights to 1.6 million acres of adjacent forest at $1 an acre. A year later Conservation International added 110,000 acres to Bolivia's Madidi National Park by a purchase of logging rights at $0.90 an acre. Both agreements were outstanding deals for the cause of conservation. Parts of the areas the two parks embrace are in the tropical Andes, an upland region extending from Venezuela west to Colombia and thence south through Ecuador and Peru to Bolivia, and comprising countless isolated ridges and valleys. The tropical Andes constitute probably the richest hotspot in the world, containing 40,000 to 50,000 plant species, or 15 to 17 percent of the world's total flora. Of these, 20,000 are found nowhere else. The environment of the region is also in parlous condition. Only about 25 percent of the forest cover is intact, and that remnant is shrinking fast.

In 1998 Suriname, the Dutch-speaking neighbor of Guyana, became the beneficiary of a one-million-dollar private gift made through Conservation International to start an offshore trust fund in aid of its forest conservation. Thus was begun the process to secure the continuous Central Suriname Nature Reserve, which at four million acres is one of the largest and probably the most pristine tropical forest in the world under protection. The trust fund, now formalized as the Suriname Conservation Foundation by the government, is receiving additional funds from Conservation International, the United Nations component of the Global Environment Facility, and the United Nations Foundation (an organization established by private funding from the American communications entrepreneur Ted Turner). The immediate goal of the trust fund managers is $15 million; by 2001 it had been more than half met. That amount, although modest by the usual standards of international aid, is expected to grow by further contributions and forest-generated income. More importantly, it was enough to persuade the government to cancel logging concessions and save for posterity its still undamaged interior wilderness.

The interplay of biology, economics, and diplomacy is high adventure of a new kind. Here, for example, is an account by Russell Mittermeier, president of Conservation International, of the genesis of the Suriname Conservation Concession (personal communication, May 15, 2001):

> Suriname has the highest percent rainforest cover of any country on Earth. In the mid-1990s, it was discovered by the Malaysian and Indonesian logging conglomerates, which were running out of land to exploit in Southeast Asia. Three companies came in and tried to secure 3 million hectares in concessions. We reacted with a strong international media campaign and some internal activity coordinated by our CI Suriname program, entirely run by Surinamers, to prevent this. One concession of 150,000 ha was granted, but the others were put on hold. Nonetheless, the threat remained, and in mid-1997 another proposal was discussed that would have put a concession immediately to the north and partly surrounding the Raleighvallen-Voltzberg Nature Reserve. This was the most important reserve in the country (and also the site of my doctoral research, so I had some vested interest). We started to discuss possible options to ensure the protection of both this reserve and the headwaters of the pristine Coppename River that flows through the reserve. Ian Bowles and I looked at the map and saw that if we tried to extend Raleighvallen south to protect the headwaters of the Coppename, we ran into another existing reserve, the Tafelberg. We then started to think big, and looked still further south to the other big protected area of the interior, the Eilerts de Haan Reserve. We came up with a number of scenarios to link these and to protect the Coppename River, and wound up with the proposal for 1.6 million hectares, a fourfold increase over the size of the three existing

reserves. We originally extended the reserve even further south, but when we consulted with the Trio Indians, with whom we had been working for 15 years, they told us that they claimed that particular piece.

In January 1998, I met with President Jules Wijdenbosch and the Minister of Natural Resources to discuss it with them. Pete [Peter Seligmann, chairman of the board], in the meantime, had secured a private commitment of $1 million to get the ball rolling, which enabled me to make a preliminary offer to the government. I told them we would jump start the process with the $1 million and that we would find additional resources later. They asked for a proposal. Over the next five months, we exchanged letters and signed an MOU [memorandum of understanding], and by June the government was ready for us to announce the creation of the reserve. This was done in a press conference in New York, with Board Member Harrison Ford in attendance and with Suriname represented by Wim Udenhout, former Suriname Ambassador to the U.S. and then an advisor to the President of Suriname. Mohammed El-Ashry, Executive Director of the GEF, also sent a letter to be read at the press conference, in which he committed his support to the project. A month later, in July, the reserve was officially declared.

Over the next two years, we worked with the GEF to finalize their commitment, we secured a contribution of $1.7 million from the UN Foundation, and we also received additional private support from the Goldman Foundation and from another private individual. In addition, we proposed the reserve to UNESCO for World Heritage status, and began with the process of creating a board for the Suriname Conservation Foundation, which was the name the Surinamers chose for the offshore trust fund. All of this went as smoothly as could be expected, and by November 2000 we were ready to officially launch the Suriname Conservation Foundation. This we did in November 2000, and we were delighted that the World Heritage Committee approved World Heritage status on the very same day that we held the first meeting of the Suriname Conservation Foundation.

While all this was happening, there was a change of government in Suriname, which was cause for some concern. The government of President Jules Wijdenbosch, which had declared the reserve, was voted out of office and he was replaced by Ronald Venetiaan, who had been president immediately prior to Wijdenbosch. There was fear that he would try to reverse the measure taken by Wijdenbosch, but this did not happen. Ambassador Wim Udenhout, who in the meantime had become our CI Program Director, worked closely with the new President. He and I met with the new President in November, and he expressed his support for the reserve. So everything seems to be progressing well.

I think that we must have set some kind of record with the Central Suriname Nature Reserve. Initial discussions began in January 1998, the reserve was declared by July 1998, World Heritage status was approved by November 2000, and the Trust Fund was created with an initial investment of $8 million also by November 2000.

The Suriname initiative illustrates the final of three stages of nature conservation. The first is the creation of individual reserves. Nowadays a special effort is made to establish biologically diverse reserves on land and in shallow marine environments, although in theory they can also be established in the open sea and on the deep ocean bottom. Reserves are the essential core of the conservation agenda but often only a rearguard action. Unless very large at the outset, they are vulnerable to human activity and the invasion of alien organisms. Even when well protected, their status is that of islands in a sea of intensifying human activity. Within these islands, separated from surrounding natural environment of the same kind, some species inevitably go extinct. The smaller the reserve, the higher the rate of extinction. So the logical second stage in a well-designed conservation program is restoration, the enlargement of reserves by encouraging the regrowth of natural habitat outward from the periphery of the core reserve, while reclaiming and restoring developed land close by to create new reserves.

The final stage of conservation is the one pioneered by Suriname with the aid of the nongovernmental organizations: to secure or rebuild wilderness by the establishment of large natural corridors that connect existing parks and reserves.

A true wilderness reserve protects in perpetuity entire faunas and floras. It shelters the largest native carnivores of the region, such as wolves, jaguars, and harpy eagles. In some cases a wilderness reserve can be made continent-wide. Such is the goal of the Wildlands Project and other of the most visionary nongovernmental organizations. Advances of such magnitude will require new levels of science, funding, and political consensus. They will become part of more sophisticated regional management, using the technology of Geographical Information Systems. In this now well-established procedure, digitized images of the distribution of habitats and species are superimposed on maps of typography, hydrology, human settlement, agriculture, industry, and transportation routes, then made part of the political process in arguing for the establishment of reserves, up to and including wildlands corridors.

The megaprojects are not a wild-eyed utopian vision. They are mainstream conservation writ large for future generations. In the Western Hemisphere they are possible as a series of long corridors pieced together from remnant natural lands that still exist from Alaska all the way to Bolivia. For North America in particular, the Wildlands Project has offered plans for a corridor to run from the Yukon to Yellowstone National Park. Another corridor is the Sky Islands Wildlands Network, which can link still wild upland habitats in New Mexico and Arizona with those in northern Mexico. A third, the Appalachian corridor, could consist of more or less continuous forest from western Pennsylvania into eastern Kentucky.

In the United States and the rest of the world the time for corridor megaprojects is now, because the windows of opportunity are closing fast.

The vision of large-scale conservation at each of the three stages is brightest, and concessions and endowments most promising, in developing countries with extensive wildlands and small populations. For example, Suriname is about the size of New York State but has (or had in 1997) only 425,000 people, 90 percent of whom live in the coastal region with about half in and around the capital city of Paramaribo. By the yardstick of international trade, concessions and endowments offer immediate long-term economic advantages to the host countries, on top of huge gains to their own and global conservation. The same calculus applies elsewhere, but the variables get tougher. In congested countries with intense competition for undeveloped land the price rises steeply, and the nongovernmental organizations find it more difficult to compete with private developers. But it can be done, given funding, public support, and ingenuity. One of the most effective means of acquisition of natural land is by purchase or gift from owners who wish to see their property kept as an inviolate reserve. The leader in this method, and consequently the foremost nongovernmental reserve manager in the world, is The Nature Conservancy.

I had a productive experience with the Conservancy in 1968, when as one of a group of young scientists I worked with TNC, then a relatively young organization, and the state of Florida to acquire Lignumvitae Key, located near the center of the Florida Keys. This small island is the site of the best old-growth lowland West Indian forest in the United States and, as it turned out, almost the entire West Indies as well. Lignumvitae Key was privately owned, and for sale. Its appearance on the market is an example of the distribution of American land-based wealth that has contributed heavily to the Conservancy's success: of the 34 million Americans who owned any part of the one billion acres of private land in 1978, the top 5 percent, or less than 1 percent of the U.S. population as a whole, owned three-fourths. There is no reason to believe that the distribution has changed greatly since then. Hence the potential transfer of large tracts of natural land from wealthier Americans by sale or gift is enormous. Reserves do not, as a rule, have to be pieced together from many small lots negotiated separately.

Among the most recent and dramatic acquisitions by The Nature Conservancy, made in November 2000, is Palmyra, an American-owned Pacific island, and one of only two atolls in the wet zone of the tropical equatorial belt that have never been inhabited in historical times. The 680 acres of islets and 15,000 acres of pristine coral reefs composing Palmyra are well worth the agreed-upon price of $37 million.

At about the same time, TNC helped to purchase and set aside a portion of the Cuatro Ciénagas (Four Marshes) in the Chihuahuan desert of north-central Mexico. The region is the site of rare spring-fed desert pools and wetlands that, because of their extreme isolation, harbor unique species of plants, invertebrates, reptiles, and fish.

Other organizations have taken the same initiative as opportunities arise. Conservation International, with a gift from Gordon Moore, a cofounder of Intel and member of the CI governing board, recently acquired a large tract of the Pantanal, a vast Everglades-like wetland that straddles the border between Brazil, Paraguay, and Bolivia, and is the largest tropical wetland in the world. The opportunities for similar acquisitions are bright throughout most of Latin America, where, as in the United States, large tracts of land are owned by a relatively small fraction of the population. The establishment of reserves becomes easier when shifts in the market make private land ownership less profitable. In the Pantanal the main source of external income has long been cattle ranching, practiced on the unflooded tracts of high ground. But profits fell off as competition increased from feedlots and ranches closer to the Brazilian processing centers and markets. Now the land can be converted into reserves at higher profit. The

CI tract is earning more money per acre from ecotourism than comparable neighboring properties are from ranching.

In Costa Rica, where land costs are low, private nature reserves have become a commonplace, established in rainforest and other natural areas by environmental NGOs or by entrepreneurs in the country's booming ecotourism business. Tourism generally has become the country's leading source of foreign revenue, exceeding even its formerly dominant banana export business.

The nongovernmental organizations, acting as the main conduit for the private sector on behalf of conservation, are distinguished from government by some of the better attributes of business corporations. They are more goal-oriented than government agencies, less rigidly bureaucratic, answerable to unpaid independent governing boards, and run by personnel evaluated at frequent intervals for the originality and quality of their work. They are opportunistic and expansive in style. They also use much the same language as corporate entrepreneurs. In acquiring environmentally valuable land, the NGO strategists analyze the needs of the "stakeholders," including indigenous people, government officials, financial sponsors of the initiative, and potential ecotourists and consumers of marketable products. They form "partnerships" with regional conservation NGOs, village councils, and philanthropic organizations. They use the partnerships to "leverage" the amount of investment in their projects and to promote the conservation ethic. Among the most effective partnerships have been those formed in various combinations between, on the one side, the World Bank, the Global Environmental Facility, and the United Nations and, on the other side, Conservation International, the IUCN – World Conservation Union, the U.S.-based World Wildlife Fund, and the international World Wide Fund for Nature. The global NGOs, unlike many large corporations, try to steer clear of local government policy and political ideology. Their focus stays on their one driving reason for existence: the protection of biological diversity.

Global NGOs are aided by the circumstance that developing countries, which harbor most of the planet's biodiversity, are also in the greatest need of economic assistance. As a result, large advances there in environmental protection can be made cost-effective with satisfactory results for all parties. The NGOs are forced to lead the way because of the lethargy of the governments of the wealthy industrialized states, whose ordinary citizens remain unconcerned about the faunas and floras of distant impoverished countries. And such being the case, the developing countries themselves see little incentive to devote their meager resources to the preservation of the natural environment, however ultimately valuable.

Yet eventually governments in both north and south will have to take over the heavy lifting from the NGOs. One recent study suggests that an investment of about $28 billion is needed to maintain at least a representative sample of Earth's ecosystems, land and sea, pole to pole. Beyond a mere sample, a comparable sum would achieve a very high yield of species-level conservation through investment in the biologically richest segments, especially in the tropics. On the one hand, according to estimates made by scientists attending the 2000 conference "Defying Nature's End," organized by Conservation International, $4 billion is needed to secure management of the approximately two million square kilometers of tropical forest wilderness now protected, at least on paper, for biodiversity and indigenous peoples, plus the acquisition and management of the remaining two million square kilometers. The result from this single investment would be a permanent wilderness belt around the equator large enough to sustain a substantial fraction of Earth's biodiversity, including the largest and most spectacular animals, such as jaguars and gorillas.

On the other hand, the hotspot areas of the world, covering less than 2 percent of Earth's land surface and serving as the exclusive home of nearly half its plant and animal species, are a more focused but difficult target. Already severely depleted in area, they are in many cases surrounded by large human populations and hence expensive to acquire and maintain. Approximately $24 billion is needed to manage in perpetuity 800,000 square kilometers already under protection, as well as to add and permanently maintain an additional 400,000 square kilometers still unprotected. By treaties, concessions, and low-invasive use of the reserves, the investments can be made attractive to the countries that govern the land.

The tropical wilderness areas and the hottest of the hotspots on the land and in shallow marine habitats, which together contain perhaps 70 percent of Earth's plant and animal species, can be saved by a single investment of roughly $30 billion. If this seems a large sum, bear in mind that it is only about one-thousandth of the annual combined gross national products of the world—or, viewed another way, one-thousandth of the value of services provided free each year by the world's natural ecosystems.

As of 2000, only about $6 billion was being allotted annually from combined governmental and private sources to sustain all of Earth's natural ecosystems. There is no reasonable prospect that NGOs can raise the amount of capital investment needed to cover in perpetuity endangered ecosystems with the greatest biodiversity. The immediate role of the NGOs is therefore that of an emergency task force: defining the problem, devising strategy, and taking local action when adequate resources can be marshaled.

Some of the needed governmental funding can be freed by ending perverse subsidies that aid individual industries but are unnecessary for the country as a whole and harmful to the environment. The world ocean fishing catch, a conspicuous example, is worth about $100 billion dockside but is sold for $80 billion, with the difference being paid by government subsidies. In the balance scales of economy and environment, the advantage to consumers is outweighed by the cost to the fishery stocks. The subsidies are one reason that all of the key ocean fisheries are now below sustainable levels. Some of their most valuable species, such as cod and haddock in the North Atlantic, have been driven to near commercial extinction—that is, to such scarcity that industries based on them have either collapsed or turned to other species. Ranching and mining also commonly benefit from perverse subsidies. In Germany government support for coal mining is so high that it would be more economical to close all the mines and send the workers home at full pay.

In an analysis published in 1998, Norman Myers and Jennifer Kent of Oxford University placed annual subsidies worldwide at $390 billion to $520 billion in agriculture, $110 billion in fossil fuels and nuclear energy, and $220 billion for water. All these and other subsidies combined exceed $2 trillion, much of which is harmful to both our economies and our governments. The average American pays $2,000 a year in subsidies, giving the lie to the belief that the American economy runs in a truly free competitive market. An additional heavy price, difficult to measure but nevertheless substantial, is levied on the natural environment, which carries the burden of extraction and consumption.

Beyond economic policy, it is to governments we must also turn for treaties that protect the global environment. The Montreal Protocol has worked to reduce and will eventually eliminate excess chlorofluorocarbons (CFCs), which thin the protective layer of ozone in the upper atmosphere. The Kyoto Protocol, if fully implemented, would slow the release of carbon dioxide and other greenhouse gases threatening to trigger the runaway warming of Earth's climate. Unfortunately, as I write in 2001, it appears as endangered as the giant panda.

Less well known are international treaties directed at the protection of biological diversity. The Convention on International Trade in Endangered Species of Wild Fauna and Flora (CITES) forbids the commercial transport across borders of live specimens and body parts of rare plants and animals. Hundreds of items are protected, from rare cacti and parrots to elephant tusks and tiger bones. Inaugurated in 1973, CITES has been effective in reducing the exploitation of rare species but remains far from perfect. The Convention on Migratory Species (CMS), operational since 1983, protects endangered migratory animals, including Siberian cranes and European bats, that cross national boundaries during their annual migrations.

The most sweeping and popular of all international treaties, however, is the Convention on Biological Diversity of the 1992 Earth Summit held in Rio de Janeiro. Now ratified by 178 countries, its provisions call for national surveys of faunas and floras, the establishment of parks and reserves, and the assessment and protection of endangered species.

The treaty-making powers of government can also be used to turn disputed territories into international peace parks. As swords can be beaten into plowshares, so battlegrounds can be turned into nature reserves. The most important potential site for such action is the demilitarized zone (DMZ) between North and South Korea. In place since the 1953 armistice agreement that ended the Korean War, the DMZ is a no-man's-land, a depopulated corridor 2.4 miles wide, winding like a ribbon for 150 miles across the Korean Peninsula. It can be converted at virtually no cost into the largest and best wildlife sanctuary of a future united Korea. Half a century of undisturbed forest growth covers its rolling hills. Signs of leopard have been spotted, and tigers may also be present. The idea of the park, first proposed by the Korean-American Ke Chung Kim, is being promoted by the DMZ Forum, an international nongovernmental group devoted exclusively to its establishment.

The strength of each country's conservation ethic is measured by the wisdom and effectiveness of its legislation in protecting biological diversity. Without dispute, the most important conservation law in the history of the United States is the Endangered Species Act. Passed in 1973 by a vote of 390–12 in the House of Representatives and 92–0 in the Senate, and signed into law by President Nixon, it was unprecedented in its sweep. Every kind of plant and animal at risk became eligible for listing. Previous legislation had only protected vertebrates, mollusks, and crustaceans. Now, under ESA provisions, the Tennessee purple cornflower, San Rafael cactus, Palos Verdes blue butterfly, and American burying beetle joined the Florida panther and golden-cheeked warbler as the legal wards of the American people. Further, in the special case of birds, mammals, and other vertebrates, not just species but local races were taken under the umbrella. (Races of invertebrates and plants remain excluded.) Finally, not just species and races on the brink of extinction but those classified as threatened—likely to become endangered—were included.

From its inception the Endangered Species Act has been burnished by its admirers, buffeted by its critics, and amended by Congress. The most important change, made in 1982, was the provision for Habitat Conservation Plans. This amendment allows landowners an "incidental take"—in other words, unintended killing—of protected plants and animals in the course of their otherwise lawful business, provided they add activities that help the affected species overall. In one such operation the International Paper Company struck a bargain with the U.S. Department of the Interior, which administers the ESA, on behalf of the red-cockaded woodpecker. This resident of southern forests has been reduced to endangered status by the cutting of large pine trees that are its exclusive nesting sites. The International Paper Company agreed to set aside a reserve within its holdings and to improve nest sites for the species, in exchange for permission to continue logging in other tracts where the woodpecker might be affected.

Given its status as a rudimentary bill of rights for biodiversity, the Endangered Species Act has been closely monitored over the years. As any conservation biologist would have predicted, the record is mixed. There have been, on the one hand, dramatic successes. The American alligator, gray whale, bald eagle, peregrine falcon, and the eastern population of the brown pelican have improved enough to be either removed from the endangered list or proposed for such delisting. On the other hand, a few, including the dusky seaside sparrow and Maryland darter, have spiraled on down to extinction. In its most recent assessment, made in 1995, the U.S. Fish and Wildlife Service, the agency within the Department of the Interior responsible for administering the act, concluded that fewer than 10 percent of the listed species were improving, while 40 percent were declining. The remainder were either stable or of unknown status.

Critics who would like to see the Endangered Species Act weakened call its imperfect record a failure. They would do as well to call a hospital emergency room a failure because more people die there than leave in good health. Better to ask for more funding and professional attention in America's natural reserves, as the public routinely does on behalf of emergency rooms.

The critics also charge that even the species saved are not worth it to the country, because implementation impedes America's economic development. Nothing could be further from the truth. At worst, the Endangered Species Act modifies development and coaxes it into new directions. It often enhances property values by recreation opportunities and other amenity benefits. Where would developers and light industry prefer to locate, next to a Douglas fir forest or a sea of Douglas fir stumps? In any case, the outright blocking of development has been a rarity. Of 98,237 projects reviewed by the federal government during interagency consultations between 1987 and 1992, only 55 were stopped cold by application of the Endangered Species Act. One of the reasons for this light touch is that endangered species tend to be concentrated in geographically limited hotspots, such as the Hawaiian rainforests and the Lake Wales Sand Ridge scrubland of central Florida. Very few are to be found in the great stretches of America's agricultural belts and ranchlands, from which so much of the anti-ESA protest arises.

At the end of the day, in a more democratic world, it will be the ethics and desires of the people, not their leaders, who give power to government and the NGOs or take it away. They will decide if there are to be more or fewer reserves, and choose whether particular species live or die. That is why I am personally encouraged by the swift growth of nongovernmental organizations devoted to conservation. The ability of people to devise initiatives fitted to the occasion, from the protection of a local riverine woodland and endangered frog species to the support of rainforest wilderness reserves and international treaties, is strong and growing. There is also a reasonable expectation that the study of biodiversity and concern for its welfare will be an increasing focus of education from kindergarten to twelfth grade and college, and beyond. What better way to teach science than to present it as a friend of life rather than as an uncontrolled destructive force?

At the risk of seeming politically correct, I will now close with a tribute to protest groups. They gather like angry bees at meetings of the World Trade Organization, the World Bank, and the World Economic Forum. They boycott insufficiently green restaurant franchises. They mass on logging roads. In response the executives and trustees they target ask, Who are these people? What are they really after? The answers to these questions are simple. They are people who feel excluded from the conference table by faceless power, and they distrust decisions secretly made that will affect their lives. They have a point. The CEOs and governing boards of the largest corporations, supported by government leaders committed to an expanding capital economy, are the commanders of the industrialized world. Like princes of old, they can, in the realm of economics at least, rule by fiat. The protesters say: Include us, and while you're at it, the rest of life.

The protest groups are the early warning system for the natural economy. They are the living world's immunological response. They ask us to listen. Julia (Butterfly) Hill, the young lady who lived 180 feet up in a California redwood tree for two years (December 1997–December 1999) in an attempt to save the redwoods of the surrounding forest, just wanted to express her opinion and change minds. Her argument was simple: it is morally wrong to cut down these ancient giants, even if you own them. She lost. She persuaded Pacific Lumber MAXXAM to save only her tree and three acres of land around it. But how many now know her name and the name of the tree (Luna), and how many know the names of the corporate executives who ordered the logging to continue around the tiny preserve?

Granted, some of the protest groups are tainted by the violent actions of a few. Rioters attacking police, burners of construction sites, and drivers of spikes into trees marked for logging deserve fines and jail terms. But the vast majority of the protesters, those honest, loud-shouting picketers dressed in turtle costumes and homeless-shelter couture, gather to demand equal time for the poor and for nature. I say bless them all. Their wisdom is deeper than their chants and tramping feet suggest, deeper than that of many of the power brokers they oppose. With the help of the media, which feeds on controversy, they keep open public discourse on crucial issues otherwise slighted. And if they are uniformly left-wing in ideology, so be it. Their youthful energies therapeutically disturb and counter the cynicism endemic to the conservative temperament.

The central problem of the new century, I have argued, is how to raise the poor to a decent standard of living worldwide while preserving as much of the rest of life as possible. Both the needy poor and vanishing biological diversity are concentrated in the developing countries. The poor, some 800 million of whom live without sanitation, clean water, and adequate food, have little chance to advance in a devastated environment. Conversely, the natural environments where most biodiversity hangs on cannot survive the press of land-hungry people with nowhere else to go.

I hope I have justified the conviction, shared by many thoughtful people from all walks of life, that the problem can be solved. Adequate resources exist. Those who control them have many reasons to achieve that goal, not least their own security. In the end, however, success or failure will come down to an ethical decision, one on which those now living will be defined and judged for all generations to come. I believe we will choose wisely. A civilization able to envision God and to embark on the colonization of space will surely find the way to save the integrity of this planet and the magnificent life it harbors.

———o———

NOTES

PROLOGUE: A LETTER TO THOREAU

xiv   On the **rise of the red maple** *(Acer rubrum)* in the eastern U.S. forests: Marc D. Abrams, *BioScience* 48 (5): 355–64 (1998).

xvi   Figures for the **high density of soil organisms** are cited, for example, by Peter M. Groffman in *Trends in Ecology & Evolution* 12 (8): 301–2 (1997); and Peter M. Groffman and Patrick J. Bohlen in *BioScience* 49 (2): 139–48 (1999).

xvii   Peter Alden of Concord, Massachusetts, organizer of the first **New England Biodiversity Day,** has compiled the list of 1,904 plant, animal, and fungal species in an unpublished report, "World's First 1000+ Species Biodiversity Day" (1998), available from Alden and, eventually, my own archives in the Library of Congress.

xvii   Henry David **Thoreau's recently published works** are *Faith in a Seed: The Dispersion of Seeds and Other Late Natural History Writings* (Washington, D.C.: Shearwater Books, Island Press, 1993) and *Wild Fruits: Thoreau's Rediscovered Last Manuscript* (New York: W. W. Norton, 2000), both edited by Bradley P. Dean.

xviii I am grateful to Stefan Cover, the leading authority on North American ants, for his suggestion that **Thoreau's ant war** was a slave raid, most likely by the red-and-brown *Formica subintegra* on the larger, black *Formica subsericea*. Both species are common in the vicinity of Walden Pond.

xix   **Thoreau's contributions to science,** including his concept of forest succession, have been well analyzed by Michael Berger in *Annals of Science* 53: 381–97 (1996), providing evidence that a longer-lived Thoreau would indeed have been viewed as a great naturalist as well as an influential pioneer ecologist.

xxii   The philosopher who described **life as a predicament** was George Santayana.

xxii   The **Abrahamic image** of the world as a source of milk and honey was taken from Aldo Leopold's *A Sand County Almanac, and Sketches Here and There* (New York: Oxford Univ. Press, 1949).

CHAPTER 1: TO THE ENDS OF EARTH

3   The **life of the McMurdo Dry Valleys** is described by John C. Priscu in *BioScience* 49 (12): 959 (1999); Ross A. Virginia and Diana H. Wall, *ibid.:* 973–83; and Diane M. McKnight et al., *ibid.:* 985–95. I am grateful to Diana Wall (personal communication) for information on the recent discovery of mites and springtails in the Dry Valleys.

4   Recent research on **life in and around Antarctic sea ice** is reported by Kathryn S. Brown in *Science* 276: 353–4 (1997); Alison Mitchell, *Nature* 387: 125 (1997); and James B. McClintock and Bill J. Baker, *American Scientist* 86 (3): 254–63 (1998).

5   On **heat-loving microbes** that live in water close to or above the boiling point, and **other extremophiles,** see Michael T. Madigan and Barry L. Marrs in *Scientific American* 276 (4): 82–7 (April 1997).

6   Research on **life in the Challenger Deep,** the ocean floor's greatest depth, is reported by Richard Monastersky in *Science News* 153 (24): 379 (1998).

6   The **extremely radiation-resistant bacterium** *Deinococcus radiodurans* is described by Patrick Huyghe in *The Sciences* 38 (4): 16–19 (July/August 1998).

7   Research on the **pariah bacteria** is described by Will Hively in *Discover* 18: 76–85 (May 1997).

7   **Deep earth SLIMEs** (subsurface lithoautotrophic microbial ecosystems): James K. Fredrickson and Tullis C. Onstott, *Scientific American* 275 (4): 68–73 (October 1996); W. S. Fyte, *Science* 273: 448 (1996); and Richard A. Kerr, *Science* 276: 703–4 (1997).

7   The search for **life on Mars and Europa:** Kathy A. Svitil, *Discover* 18: 86–8 (May 1997); Richard A. Kerr, *Science* 277: 764–5 (1997); Michael H. Carr et al., *Nature* 391: 363–5 (1998); Robert T. Pappalardo, James W. Head, and Ronald Greeley, *Scientific American* 281 (4): 54–63 (October 1999); Christopher F. Chyba, *Nature* 403: 381–2 (2000). I am grateful to Matthew J. Holman for basic information on the internal heat of Mars and for directing me to the key recent model by F. Sohl and T. Spohn, *Journal of Geophysical Research* 102 (E1): 1613–35 (1997).

9   **Life in Lake Vostok, Antarctica:** Warwick F. Vincent, *Science* 286:

2094–5 (1999); and Frank D. Carsey and Joan C. Horvath, *Scientific American* 281 (4): 62 (October 1999).

9   On the independent flora and fauna of **Movile Cave, Romania:** E. Skindrud in *Science News* 149: 405 (1996); and on the independent biota of **Cueva de Villa Luz:** Charles Petit in *U.S. News & World Report* 124 (5): 59–60 (February 9, 1998).

11   Some of the scientific evidence pertaining to **Gaia** is evaluated (favorably) by Jim Harris and Tom Wakeford in *Trends in Ecology & Evolution* 11 (8): 315–16 (1996); and David M. Wilkinson, *ibid.*, 14 (7): 256–7 (1999). Updates on Gaia research can be followed in *Gaia Circular* (Newsletter of the Society for Research and Education in the Earth System Science). James Lovelock, who invented the concept, details its history in his memoir *Homage to Gaia: The Life of an Independent Scientist* (New York: Oxford Univ. Press, 2000).

12   For a fuller description of the **principles of classification,** together with an account of the evolutionary origin of species, see Edward O. Wilson, *The Diversity of Life* (Cambridge, MA: Belknap Press of Harvard Univ. Press, 1992).

14   On the **superabundant oceanic bacterium *Prochlorococcus:*** Sallie W. Chisholm et al., *Nature* 334: 340–3 (1988); and Conrad W. Mullineaux, *Science* 283: 801–2 (1999).

14   On **invisible organisms and dark matter** in the open ocean: Farooq Azam, *Science* 280: 694–6 (1998).

15   Research on the **diversity of fungi** is reported by Robert M. May in *Nature* 352: 475–6 (1991); and Gilbert Chin in *Science* 289: 833 (2000).

15   The **diversity of nematodes:** Claus Nielsen in *Nature* 392: 25–6 (1998); and Tom Bongers and Howard Ferris, *Trends in Ecology & Evolution* 14 (6): 224–8 (1999).

15   The **new lobster-inhabiting phylum Cycliophora:** Simon Conway Morris, *Nature* 378: 661–2 (1995); and Peter Funch and Reinhardt M. Kristensen, *Nature* 378: 711–4 (1995).

16   The **phyla of invertebrate animals** are defined and described in detail by Richard C. Brusca and Gary J. Brusca in their standard textbook *Invertebrates* (Sunderland, MA: Sinauer Associates, 1990).

17   The **continuing discovery of U.S. and Canadian flowering plants** is reported by Susan Milius in *Science News* 155 (1): 8–10 (1999).

17   On **diversity in the amphibians:** James Hanken, *Trends in Ecology & Evolution* 14 (1): 7–8 (1999), and personal communication.

17   On the **discovery of new kinds of mammals:** Bruce D. Patterson in *Biodiversity Letters* 2 (3): 79–86 (1994); and Virginia Morrell in *Science* 273: 1491 (1996).

18   The number of **new species of monkeys and other primates** was
      provided by one of their principal discoverers, Russell A. Mittermeier
      (personal communication).

18   The **saola and other new large mammals** from Vietnam: Alan Rabi-
      nowitz in *Natural History* 106 (3): 14–18 (April 1997); John Whitfield in
      Nature 396: 410 (1998); and Daniel Drollette in *The Sciences* 40 (1): 16–19
      (January/February 2000).

18   The **number of bird species** and potential for new ones: Trevor Price in
      Trends in Ecology & Evolution 11 (8): 314–15 (1996).

20   The **record number of tree species,** established in Bahia State, Brazil, by
      a team from the New York Botanical Garden, is reported by James Brooke
      in the Environment Section of the *New York Times* (March 30, 1993); and
      the butterfly record, set by Gerardo Lamas, Robert K. Robbins, and Don-
      ald J. Harvey, is reported in *Publicaciones del Museo de Historia Natural,
      Universidad Nacional Mayor de San Marcos* (Ser. A: Zoologia) 40: 1–19
      (1991).

20   The **record number of vine and epiphytic** species on a single tree (in
      New Zealand) is reported by K. J. M. Dickinson, A. F. Mark, and
      B. Dawkins in the *Journal of Biogeography* 20: 687–705 (1993).

20   On **mouth-dwelling bacteria:** Jane Ellen Stevens, *BioScience* 46 (5):
      314–17 (1996).

CHAPTER 2: THE BOTTLENECK

22   The concept of directions of the **twentieth and twenty-first centuries** as
      presented here is modified and described in slightly altered form from my
      article of that title in *Foreign Policy* 119: 34–5 (summer 2000).

23   The **ecological footprint** of human impact on the environment was
      devised by William E. Rees and Mathis Wackernagel in AnnMari Jansson
      et al., eds., *Investing in Natural Capital: The Ecological Economics Approach
      to Sustainability* (Washington, D.C.: Island Press, 1994), pp. 362–90; and
      updated in a personal communication from Mathis Wackernagel (January
      24, 2000) (Redefining Progress, 1 Kearny St., San Francisco, CA); and by
      Wackernagel et al. in *Living Planet Report 2000* (Gland, Switzerland:
      World Wide Fund for Nature, 2000), pp. 10–12.

23   The data and arguments in the **dialogue between economist and ecolo-
      gist** were drawn from many sources. Among the most up-to-date were the
      series *Living Planet Report* (1998 and 1999) of the World Wide Fund for
      Nature (Gland, Switzerland), the New Economics Foundation (London),
      and the World Conservation Monitoring Center (Cambridge, England);
      and *World Resources 2000–2001: People and Ecosystems—The Fraying Web of
      Life,* produced by the World Resources Institute in collaboration with
      the United Nations Development and Environment Programmes and the

World Bank (Oxford: Elsevier Science, 2000; Washington, D.C.: World Resources Institute, 2000; summary available at www.elsevier.com/locate/worldresources).

28   Among the key sources I used for the account of **human population growth** were *How Many People Can the Earth Support?*, by Joel E. Cohen (New York: W. W. Norton, 1995); "Population Policy: Consensus and Challenges," by Lori S. Ashford and Jeanne A. Noble, *Consequences* (Saginaw Valley State Univ., University Center, MI) 2 (2): 25–35 (1996); *Beyond Malthus: Sixteen Dimensions of the Population Problem* (Worldwatch Paper 143), by Lester R. Brown, Gary Gardner, and Brian Halweil (Washington, D.C.: Worldwatch Institute, 1998); and *Global Environmental Outlook 2000* (United Nations Environment Programme) (London: Earthscan Publications, 1999). The projections to 2050 are in part from *World Population Prospects: The 1998 Revision, Volume 1: Comprehensive Tables* (New York: United Nations Publication, Sales No. E.19.XII.9, 1999).

32   The **percentages of countries with 40 percent or more of the population under fifteen years of age** are from *The New York Times 1999 World Almanac.*

33   On the **number of East Indians and Americans supportable by the world grain harvest:** Lester R. Brown et al., *Beyond Malthus: Sixteen Dimensions of the Population Problem* (Worldwatch Paper 143).

34   The **absolute limit of 17 billion people** imposed by the ultimate photosynthetic origin of food was calculated by John M. Gowdy and Carl N. McDaniel and reported in *Ecological Economics* (Journal of the International Society for Ecological Economics, Amsterdam, the Netherlands) 15 (3): 181–92 (1995).

34   The reach of **type I and type II civilizations in space** are detailed by Ian Crawford in *Scientific American* 283 (1): 38–43 (July 2000).

34   The summary of **China's water resources and agricultural potential** is drawn mostly from the MEDEA Special Study, "China Agriculture: Cultivated Land Area, Grain Projections, and Implications," a 1997 report to the National Intelligence Council, the umbrella group of U.S. intelligence organizations. I have also used information on Chinese water resources provided by Sandra Postel in *Pillar of Sand: Can the Irrigation Miracle Last?* (New York: W. W. Norton, 1999). I am grateful to Michael B. McElroy, one of the authors of the MEDEA report, for information on Chinese policy after 1997.

CHAPTER 3: NATURE'S LAST STAND

42   The **Living Planet Index** is presented and documented in the annual *Living Planet Report* (1998–2000) of the World Wide Fund for Nature, the New Economics Foundation, and the World Conservation Monitoring

Centre (Gland, Switzerland: World Wide Fund for Nature). Its evaluation of the state of the natural environment is upheld by a contemporary report in *World Resources 2000–2001: People and Ecosystems—The Fraying Web of Life*, produced by the World Resources Institute in collaboration with the United Nations Development and Environment Programmes and the World Bank (Oxford: Elsevier Science, 2000; Washington, D.C.: World Resources Institute, 2000; summary available at www.elsevier.com/locate/worldresources).

43   The story of the **Hawaiian fauna and flora** has been drawn from many sources, including those summarized by the present author in *The Diversity of Life* (Cambridge, MA: Belknap Press of Harvard Univ. Press, 1992); Elizabeth Royte in *National Geographic* 188 (3): 4–37 (September 1995); Lucius G. Eldredge and Scott E. Miller in *Bishop Museum Occasional Papers* (Honolulu) 48: 3–22 (1997); James K. Liebherr and Dan A. Polhemus in *Pacific Science* 51 (4): 490–504 (1997); Stuart L. Pimm, Michael P. Moulton, and Lenora J. Justice in *Philosophical Transactions of the Royal Society of London* (Ser. B: Biological Sciences) 344 (1307): 27–33 (1994); L. G. Eldredge and S. E. Miller in *Bishop Museum Occasional Papers* (Honolulu) 55: 3–15 (1998); Warren L. Wagner et al., *ibid.* 60: 1–58 (1999); and George W. Staples et al., *ibid.* 65: 1–35 (2000).

50   The new discipline of **conservation biology** is well summarized by Richard B. Primack in *A Primer of Conservation Biology*, second edition (Sunderland, MA: Sinauer Associates, 2000). Among many scientific journals devoted to the subject, the broadest and most representative is *Conservation Biology*, published by Blackwell Science (Boston, MA) for the International Society of Conservation Biology.

51   The story of the highly endangered **Vancouver Island marmot** is told by publications of the Marmot Recovery Foundation (Vancouver, British Columbia, www.marmots.org) in collaboration with World Wildlife Fund Canada. I am grateful to Andrew A. Bryant, the principal ecologist working on the species, for details concerning its current status (personal communication).

53   The **destruction of the native tree snails** of Hawaii and the Society Islands is described in the IUCN Invertebrate Red Data Book (1983); and in more detail by James Murray et al. in *Pacific Science* 42 (3,4): 150–3 (1988); and Nancy B. Benton et al., *America's Least Wanted* (Arlington, VA: The Nature Conservancy, 1996). I am grateful to Bryan C. Clarke and Werner Loher (personal communication) for additional information on the fate of the Moorean *Partulina* snails.

54   The definitive work on the **decline of frogs and other amphibians** is the analysis by Jeff E. Houlahan et al. presented in *Nature* 404: 752–5 (2000), as a report on data supplied by 200 biologists from 936 populations in

30 countries, mostly European and North American. A parallel **decline of reptiles** is reported by J. Whitfield Gibbons et al. in *BioScience* 50 (8): 653–66 (2000).

56 The **role of inbreeding** in the decline of species is evaluated in the greater prairie chicken by Ronald L. Westemeier et al., *Science* 282: 1695–8 (1998); the Glanville fritillary by Ilik Saccheri et al., *Nature* 392: 491–4 (1998); and in the cheetah by T. M. Caro and M. Karen Laurenson, *Science* 263: 485–6 (1994).

57 The demise of the wild populations of **Schaus' swallowtail** during Hurricane Andrew is recounted by Michael J. Bean in *Wings* (Xerces Society, Portland, OR) 17 (2): 12–15 (1993).

58 The **Centinela catastrophe,** in which scores of Ecuadorian plant species were extinguished during the 1980s, is described by E. O. Wilson in *The Diversity of Life* (Cambridge, MA: Belknap Press of Harvard Univ. Press, 1992).

58 The decline of the rich **freshwater mussel fauna** of the United States is described by William Stolzenburg in *Nature Conservancy,* pp. 17–23 (November/December 1992). The devastating effects of impoundments on the thirty species of Alabama's Black Warrior and Tombigbee Rivers are documented by James D. Williams et al. in the *Bulletin of the Alabama Museum of Natural History* 13: 1–10 (1992).

58 The **loss of habitat,** especially forest cover, in the United States, including some data on other parts of the world, is detailed by Reed F. Noss and Robert L. Peters in *Endangered Ecosystems: A Status Report of America's Vanishing Habitat and Wildlife* (Washington, D.C.: Defenders of Wildlife, 1995); R. L. Peters and R. F. Noss, *Defenders,* pp. 16–27 (Fall 1995); and Reed F. Noss, Edward T. LaRoe III, and J. Michael Scott, *Endangered Ecosystems of the United States: A Preliminary Assessment of Loss and Degradation* (Washington, D.C.: U.S. Department of the Interior, National Biological Service, 1995).

59 The **decline of mammal species** in the national parks of North America is documented by William D. Newmark in *Conservation Biology* 9 (3): 512–26 (1995).

59 The condition of the **world's tropical forests,** with special reference to the Amazon rainforest, was drawn from multiple sources, including the following: *Living Planet Report 1998* (Gland, Switzerland: World Wide Fund for Nature, 1998); William F. Laurance et al., *Ecology* 79 (6): 2032–40 (1998); W. F. Laurance, *Natural History* 107 (6): 34–51 (July/August 1998); Nick Brown, *Trends in Ecology & Evolution* 13 (1): 41 (1998); Emil Salim and Ola Ullsten, cochairs, *Our Forests, Our Future* (Report of the World Commission on Forests and Sustainable Development) (Cambridge, UK: Cambridge Univ. Press, 1999); Claude Gascon, G. Bruce Williamson, and

Gustavo A. B. da Fonseca, *Science* 288: 1356–8 (2000); Bernice Wuethrich, *Science* 289: 35–7 (2000); and William F. Laurance et al., *Science* 291: 438–9 (2001). For additional information and advice on tropical deforestation, including the most recent satellite data, I am grateful to Claude Gascon, Richard A. Houghton, Norman Myers, and Marc Steininger.

60 The **global hotspots** of the world, habitats and regions of habitats with dense concentrations of species found nowhere else and in many cases endangered, were first clearly delineated by Norman Myers in *The Environmentalist* 8 (3): 187–208 (1988) and *ibid.* 10 (4): 243–56 (1990); and updated and beautifully illustrated by Russell A. Mittermeier, Norman Myers, et al. in *Hotspots: Earth's Biologically Richest and Most Endangered Terrestrial Ecoregions* (Mexico City: CEMEX, Conservation International, 1999). See also a recent summary by Norman Myers et al. in *Nature* 403: 853–8 (2000).

67 On the measurements and predicted biological effects of **global warming:** Walter V. Reid and Mark C. Trexler, *Drowning the National Heritage: Climate Change and U.S. Coastal Biodiversity* (Washington, D.C.: World Resources Institute, 1991); Robert L. Peters and Thomas E. Lovejoy, eds., *Global Warming and Biological Diversity* (New Haven: Yale Univ. Press, 1992); E. O. Wilson, *The Diversity of Life* (Cambridge, MA: Belknap Press of Harvard Univ. Press, 1992); Christopher B. Field et al., *Confronting Climate Change in California: Ecological Impacts on the Golden State* (Cambridge, MA: Union of Concerned Scientists Publications, 1999); Richard Monastersky, *Science News* 156 (9): 136–8 (1999). The 2001 projections by the Intergovernmental Panel on Climate Change (IPCC) of global warming are reported in *Science* 291: 566 (2001) by Richard A. Kerr. I have also consulted the IPCC Groups I and II summaries for policy makers, and am grateful to James J. McCarthy, one of the group cochairs, for reviewing my brief account of the IPCC projections.

70 The identity and history of **invasive species,** especially in the United States, have been documented in a series of excellent reports and popular books. As a source for the present account, they include David Pimental et al., *BioScience* 50 (1): 53–65 (2000); Walter E. Parham, *Harmful Nonindigenous Species in the United States* (Washington, D.C.: Office of Technology Assessment, Congress of the United States, 1993); Corinna Gilfillan et al., *Exotic Pests* (Washington, D.C.: National Audubon Society, 1994); Stuart Pimm, *The Sciences* 34 (3): 16–19 (May/June 1994); Bruce A. Stein and Stephanie R. Flack, eds., *America's Least Wanted* (Arlington, VA: The Nature Conservancy, 1996); Donald R. Strong and Robert W. Pemberton, *Science* 288: 1969–70 (2000); Bill N. McKnight, ed., *Biological Pollution: The Control and Impact of Invasive Exotic Species* (Indianapolis: Indiana

Academy of Sciences, 1993); Daniel Simberloff, Don C. Schmitz, and Tom C. Brown, eds., *Strangers in Paradise: Impact and Management of Nonindigenous Species in Florida* (Washington, D.C.: Island Press, 1997); and Chris Bright, *Life Out of Bounds: Bioinvasion in a Borderless World* (New York: W. W. Norton, 1998). An account of the introduction of the European starling into the United States is given by Anthony C. Janetos in *Consequences* (Saginaw Valley State Univ., University Center, MI) 3 (1): 17–26 (1997).

CHAPTER 4: THE PLANETARY KILLER

79   The state of the **Sumatran rhinoceros:** Ronald M. Nowak, *Walker's Mammals of the World,* Volume II, fifth edition (Baltimore, MD: Johns Hopkins Univ. Press, 1991); and Mark Cherrington, *The Sciences* 38 (1): 15–17 (January/February 1998). I am also grateful for the expert counsel of William Conway, Alan Rabinowitz, Edward Maruska, Terri Roth, and Thomas Foose.

83   The **return of the California condor** is chronicled by Joanna Behrens and John Brooks in *Endangered Species Bulletin* 25 (3): 8–9 (2000).

84   The recovery of the **Mauritian kestrel** is detailed by David Quammen in *The Song of the Dodo: Island Biogeography in an Age of Extinctions* (New York: Scribner, 1996). Its genetic impoverishment is described by Jim J. Groombridge et al., *Nature* 403: 616 (2000).

88   The decline of the **Tibetan antelope,** widely noted by conservation organizations, was described by Marion Lloyd in the *Boston Globe,* p. 1 (March 15, 2000). That of the **white abalone** is analyzed by Mia J. Tegner, Lawrence V. Basch, and Paul K. Dayton in *Trends in Ecology & Evolution* 11 (7): 278–80 (1996).

89   The critically **endangered tree species** of the world were censused by the World Conservation Monitoring Centre, as reported by Nigel Williams in *Science* 281: 1426 (1998). The status of those on **Juan Fernández** are reviewed by Tod F. Stuessy et al. in *Rare, Threatened, and Endangered Flora of Asia and the Pacific Rim* (Monograph Series No. 16), Ching-I. Peng and Porter P. Lowry II, eds. (Taipei: Academia Sinica, 1998), pp. 243–57.

90   The fate of the **Hawaiian po'ouli** is described by Stuart L. Pimm, Michael P. Moulton, and Lenora J. Justice in *Philosophical Transactions of the Royal Society of London* (Ser. B: Biological Sciences) 344 (1307): 27–33 (1994).

91   The decline of **Australian native mammals** is reviewed by Christopher John Humphries and Clemency Thorne Fisher in *Philosophical Transactions of the Royal Society of London* (Ser. B: Biological Sciences) 344 (1307): 3–9 (1994), and Timothy F. Flannery in *Science* 283: 182–3 (1999). The census of endangered species is provided in the *1996 IUCN Red List of*

Threatened Animals, compiled and edited by Jonathan Baillie and Brian Groombridge (Gland, Switzerland: IUCN Species Survival Commission, 1996).

92   Among many recent articles and books on the **fauna of Madagascar,** one of the best, most thorough, and up-to-date is *The Eighth Continent: Life, Death, and Discovery in the Lost World of Madagascar,* by Peter Tyson (New York: William Morrow, 2000).

94   Definitive accounts of the extinction of **New Zealand birds,** including especially the moas, are given by Atholl Anderson in *Prodigious Birds: Moas and Moa-hunting in Prehistoric New Zealand* (New York: Cambridge Univ. Press, 1989); Alan Cooper et al., *Trends in Ecology & Evolution* 8 (12): 433–7 (1993); Jared Diamond, *Science* 287: 2170–1 (2000); and R. N. Holdaway and C. Jacomb, *Science* 287: 2250–4 (2000).

95   Accounts of the decline of **Polynesian birds** are documented by Storrs L. Olson and Helen F. James in their pioneering descriptions of new fossil species (*Ornithological Monographs* nos. 45 and 46) (Washington, D.C.: American Ornithologists' Union, 1991), 88 pp. General accounts are provided by Tom Dye and David W. Steadman in *American Scientist* 78: 207–15 (1990) and by Stuart L. Pimm, Michael P. Moulton, and Lenora J. Justice, *Philosophical Transactions of the Royal Society of London* (Ser. B: Biological Sciences) 344 (1307): 27–33 (1994).

96   The **concept of filtering** in mass extinctions was developed by Stuart L. Pimm et al., *ibid.,* and by Andrew Balmford, *Trends in Ecology & Evolution* 11 (5): 193–6 (1996).

97   The **harvesting of animals in the Middle and Upper Paleolithic periods** in the Mediterranean area is analyzed by Mary C. Stiner et al. in *Science* 283: 190–4 (1999). The movements of people and farming through Neolithic Europe are described by Luigi L. Cavalli-Sforza, *Genes, Peoples, and Languages,* trans. Mark Seielstad (New York: North Point Press, 2000).

99   **Extinction rates and longevity of species** through geological time are reviewed by multiple authors in Edward O. Wilson and Frances M. Peter, eds., *BioDiversity* (Washington, D.C.: National Academy Press, 1988) and E. O. Wilson, *The Diversity of Life* (Cambridge, MA: Belknap Press of Harvard Univ. Press, 1992).

99   The various methods of **estimating extinction rates** are variously reviewed by Georgina M. Mace and Russell Lande in *Conservation Biology* 5 (2): 148–57 (1991); E. O. Wilson in *The Diversity of Life* (Cambridge, MA: Belknap Press of Harvard Univ. Press, 1992); and articles in *Philosophical Transactions of the Royal Society* (Ser. B: Biological Sciences) 344 (1307) (1994), revised and updated as the monograph *Extinction Rates,*

John H. Lawton and Robert M. May, eds. (New York: Oxford Univ. Press, 1995).

CHAPTER 5: HOW MUCH IS THE BIOSPHERE WORTH?

103　On the **ivory-billed woodpecker:** Alexander Wilson is quoted from page 20 of his *American Ornithology; or the Natural History of the Birds of the United States* (Philadelphia: Bradford and Inskeep, 1808–14); Roger Tory Peterson's *A Field Guide to the Birds,* revised (Boston: Houghton Mifflin, 1934) is the source of information of the species' range in the 1930s, with the descent toward extinction charted in subsequent editions. Occasional reports of American ivorybill sightings still trickle in, but have never been confirmed. One from the Pearl River Forest north of New Orleans in 2000 was detailed enough to excite birders, but, again, subsequent searches came up empty (*Boston Globe,* p. 2, November 11, 2000).

105　The economic **value of the world's ecosystems** was estimated by a team of Robert Costanza and twelve other scientists and economists in *Nature* 387: 253–60 (1997).

106　The definitive general review of **ecosystems services,** by thirty-two experts in different topics, is *Nature's Services: Societal Dependence on Natural Ecosystems,* Gretchen C. Daily, ed. (Washington, D.C.: Island Press, 1997).

107　The **economic value of forests** in the Catskill Mountains and plantings to replace **Atlanta's missing trees** and the effect on rainwater runoff are cited by Peter H. Raven et al. in *Teaming with Life: Investing in Science to Understand and Use America's Living Capital* (Washington, D.C.: The President's Committee of Advisors on Science and Technology [PCAST], Biodiversity and Ecosystems Panel, 1999). The estimates were made by the nongovernmental organization American Forests using formulas developed by the Natural Resource Conservation Service.

108　The key issues of **biodiversity, ecosystems stability, and ecosystems productivity** were reviewed recently by David Tilman in *Ecology* 80 (5): 1455–74 (1999) and *Nature* 405: 208–11 (2000); Kevin S. McCann in *Nature* 405: 228–33 (2000); Jocelyn Kaiser in *Science* 289: 1282–3 (2000); and F. Stuart Chapin III et al. in *Nature* 405: 234–42 (2000). For topics in mathematical theory, see Michael Loreau in *Proceedings of the National Academy of Sciences, USA* 95 (10): 5632–6 (1998); and Felix Schläpfer, Bernhard Schmid, and Irmi Seidl, *Oikos* 84 (2): 346–52 (1999). The role of microorganism biodiversity in fresh-water environments is analyzed by Robert G. Wetzel in the *Archiv für Hydrobiologie:* Special Issues: *Ergebnisse der Limnologie* (Advances in Limnology) 54: 19–32 (1999). The concept of organisms as ecosystems engineers is developed, with many examples, by Clive G. Jones, John H. Lawton, and Moshe Shachak, *Oikos* 69 (3): 373–86 (1994).

113   The **blue whale** economic analysis by Colin W. Clark is in the *Journal of Political Economy* 81 (4): 950–61 (1973). This and other examples of the weakness of purely econometric valuation are given by David Ehrenfeld in *Beginning Again: People and Nature in the New Millennium* (New York: Oxford Univ. Press, 1993).

114   On the slightly more than one hundred **plant species that supply the world with food:** Robert and Christine Prescott-Allen, *Conservation Biology* 4 (4): 365–74 (1990). The estimate is based on data from the 146 countries from which the Food and Agriculture Organization of the United Nations (FAO) collected information.

114   On **potential new crop species:** E. O. Wilson, *The Diversity of Life* (Cambridge, MA: Belknap Press of Harvard Univ. Press, 1992).

114   On **new strains and genes for existing crops:** Erich Hoyt, *Conserving the Wild Relatives of Crops* (Gland, Switzerland: World Conservation Organization, International Board for Plant Genetic Resources, and World Wide Fund for Nature, 1988).

114   The **genetic engineering of crops,** because of its importance and attending controversy, has spawned a very large literature in a short period of time. Some of the sources I have used in preparing my brief review include the following. On the potential benefits of engineering: Charles C. Mann and Dennis Normile, *Science* 283: 310–6 (1999); Mary Lou Guerinot, *Science* 287: 241, 243 (2000); Elizabeth Pennisi, *Science* 288: 2304–7 (2000); Anne Simon Moffat, *Science* 290: 253–4 (2000); Michelle Marvier, *American Scientist* 89: 160–7 (2001); J. Madeleine Nash and Simon Robinson, *Time* 156(5): 38–46 (July 31, 2000). On risk, and controversy: Dean D. Metcalfe et al., *Critical Reviews and Food Science and Nutrition* 36(S): S165–86 (1996); Issue Paper, Council for Agricultural Science and Technology No. 12 (1999), 8 pp.; Joy Bergel-son, Colin B. Purrington, and Gale Wichmann, *Nature* 395: 25 (1998); Tanja H. Schuler et al., *Trends in Biotechnology* 17: 210–6 (1999); News and Editorial Staffs, *Science* 286: 2243 (1999); Dennis Avery, *World Link,* pp. 8–9 (July/August 1999); Adrian Murdoch, interviewing Chad Holliday, *World Link,* pp. 36–9 (November/December 1999); Norman C. Ellstrand, Honor C. Prentice, and James F. Hancock, *Annual Review of Ecology and Systematics* 30: 539–63 (1999); Jill Rubin, *Masspirg* (Massachusetts Public Interest Research Group) 18 (3): 4–5 (2000); Klaus M. Leisinger, *Foreign Policy* 119: pp. 113–22 (summer 2000); Miguel A. Altieri, *Foreign Policy* 119: pp. 123–31 (summer 2000); Rosie S. Hails, *Trends in Ecology & Evolution* 15 (1): 14–8 (2000); A. R. Watkinson et al., *Science* 289: 1554–7 (2000). On

compromise, treaty, and regulation: Royal Society of London, U.S. National Academy of Sciences, Brazilian Academy of Sciences, Chinese Academy of Sciences, Indian National Science Academy, Mexican Academy of Sciences, and Third World Academy of Sciences, *Transgenic Plants and World Agriculture* (Washington, D.C.: National Academy Press, 2000); Cyril Kormos and Layla Hughes, *Regulating Genetically Modified Organisms: Striking a Balance Between Progress and Safety* (Washington, D.C.: Center for Applied Biodiversity Science, Conservation International, 2000); Colin Macilwain, *Nature* 404: 693 (2000); Richard J. Mahoney, *Science* 288: 615 (2000); Tim Beardsley, *Scientific American* 282 (4): 42–3 (April 2000). I am grateful to Thomas E. Nickson and Jerry J. Hjelle of Monsanto for frank discussions of their company's engagement and the advantages and risks of transgenic crops.

118   The expression **"evergreen revolution"** was first used by the Indian agricultural scientist M. S. Swaminathan in the mid-1990s; see, for example, his *Sustainable Agriculture: Towards an Evergreen Revolution* (Delhi, India: Konark Pvt. Ltd., 1996).

118   The estimates of the **contribution of wild species to current pharmaceuticals,** and their consequent commercial value, are from Douglas J. Futuyma, *Science* 267: 41–2 (1995); E. O. Wilson, *The Diversity of Life* (Cambridge, MA: Belknap Press of Harvard Univ. Press, 1992); Peter H. Raven et al. *Teaming with Life* (Washington, D.C.: The President's Committee on Science and Technology, 1999); and Colin Macilwain, *Nature* 392: 535–40 (1998).

121   The structure and biochemical function of **cyclosporin,** the immune blocking agent discovered in a fungus, is described by Christopher T. Walsh, Lynne D. Zydowsky, and Frank D. McKeon in *The Journal of Biological Chemistry* 267 (19): 13115–18 (July 5, 1992); and Stuart L. Schreiber and Gerald R. Crabtree, *Immunology Today* 13 (4): 136–42 (1992), and *The Harvey Lectures,* series 91, pp. 99–114 (1997).

121   The account here of the **discovery of the painkiller epibatidine** from dendrobatid frog poison is taken from accounts by David Bradley, *Science* 261: 1117 (1993); Charles W. Myers and John W. Daly, *Science* 262: 1193 (1993); and, especially, Mark J. Plotkin, *Medicine Quest: In Search of Nature's Healing Secrets* (New York: Viking Penguin, 2000).

123   The story of the Bornean *Callophylum* and the **discovery of the AIDS suppressant (+)-calanolide** is drawn from an account by Robert Cook in the "Arnold Arboretum of Harvard University" supplement of the *Harvard University Gazette,* pp. 1, 4 (November 1996). The drug is in an advanced stage in anti-HIV testing by Sarawak MediChem Pharmaceuticals, Inc.

125 Plants used in **traditional medicine** are described by James L. Castner, Stephen L. Timme, and James A. Duke in *A Field Guide to Medicinal and Useful Plants of the Upper Amazon* (Gainesville, FL: Feline Press, 1998).

126 On the **extraction of pharmaceuticals** from rainforest: Michael J. Balick and Robert Mendelsohn, *Conservation Biology* 6 (1): 128–30 (1992).

126 The **Petén rainforest industry** was reported by Laura Tangley in *U.S. News & World Report* 124 (15): 40–1, 44 (April 20, 1998).

127 The roles of **Cetus Corporation and Yellowstone National Park** in the development of polymerase chain reaction (PCR) are traced by William B. Hull in *Biodiversity* (Consultative Group on Biological Diversity) 8 (1): 1–2 (1998). I have drawn accounts of other bioprospecting efforts variously from Leslie Roberts, *Science* 256: 1142–3 (1992); Andrew Pollack in the *New York Times*, p. C10 (March 5, 1992); Ricardo Bonalume Neto and David Dickson, *Nature* 400: 302 (1999); Hunter Jackson, NPS Pharmaceuticals, personal communication (May 27, 1993); and Daniel H. Janzen, personal communication, updating details of the INBio-Merck agreement.

CHAPTER 6: FOR THE LOVE OF LIFE

130 I presented the **reasons for doubting that an ecosystem can be rebuilt** from the ground up, microorganisms and all, in *Consilience: The Unity of Knowledge* (New York: Knopf, 1998).

130 **Environmental ethics** is a large subject, addressed by a small academic industry and unfortunately largely ignored by scholars in other fields and by the public. A good reading list would include Aldo Leopold, *A Sand County Almanac, and Sketches Here and There* (New York: Oxford Univ. Press, 1949), and *For the Health of the Land* (Washington, D.C.: Island Press/Shearwater Books, 1999); Holmes Rolston III, *Philosophy Gone Wild: Essays in Environmental Ethics* (Buffalo, NY: Prometheus Books, 1986); Bill McKibben, *The End of Nature* (New York: Random House, 1989); Steven C. Rockefeller and John C. Elder, eds., *Spirit and Nature: Why the Environment Is a Religious Issue* (Boston, MA: Beacon Press, 1992); David R. Brower, with Steve Chapple, *Let the Mountains Talk, Let the Rivers Run: A Call to Those Who Would Save the Earth* (San Francisco: HarperCollins, 1995); Theodore Roszak, Mary E. Gomes, and Allen D. Kanner, *Ecopsychology: Restoring the Earth, Healing the Mind* (San Francisco: Sierra Club Books, 1995); Philip Shabecoff, *A New Name for Peace: International Environmentalism, Sustainable Development and Democracy* (Hanover, NH: Univ. Press of New England, 1996); Stephen R. Kellert, *Kinship to Mastery: Biophilia in Human Evolution and Development* (Washington, D.C.: Island Press, 1997); Daniel C. Maguire and Larry L. Rasmussen, eds., *Ethics for a Small Planet: New Horizons on Population, Consumption, and Ecology* (Albany, NY: State Univ. of New York Press,

1998); Thomas Berry, *The Great Work: Our Way into the Future* (New York: Bell Tower, 1999); James Eggert, *Song of the Meadowlark: Exploring Values for a Sustainable Future* (Berkeley, CA: Ten Speed Press, 1999); Martin Gorke, *Artensterben: Von der ökologischen Theorie zum Eigenwert der Natur* (Stuttgart, Germany: Klett-Cotta, 1999). There is in addition a professional journal, *Environmental Ethics,* published by the Center for Environmental Philosophy and the University of North Texas, Denton, Texas.

133   I owe the phrase **"investment in immortality"** to Kenneth Small, *Politics and the Life Sciences* 16 (2): 183–92 (1997).

133   The changing sign in the **Rocky Mountain campsite** was described by Holmes Rolston III in *Garden* 11 (4): 2–4, 31–2 (July/August 1987).

134   I introduced the term and **concept of biophilia** in *Biophilia* (Cambridge, MA: Harvard Univ. Press, 1984). The idea has been extended by many writers, including the multiple authors in Stephen R. Kellert and Edward O. Wilson, eds., *The Biophilia Hypothesis* (Washington, D.C.: Island Press/Shearwater Books, 1993); and Stephen R. Kellert, *Kinship to Mastery* (Washington, D.C.: Island Press, 1997).

134   Building on the data and concepts of Jay Appleton (*The Experience of Landscape;* New York: Wiley, 1975) and others, Gordon H. Orians introduced the idea of the **environment hereditarily preferred by humans** in J. S. Lockard, ed., *The Evolution of Human Social Behavior* (New York: Elsevier, 1980). It has been developed and documented further by Orians and Judith H. Heerwagen in J. Barkow, Leda Cosmides, and John Tooby, eds., *The Adapted Mind: Evolutionary Psychology and the Generation of Culture* (New York: Oxford Univ. Press, 1992); Heerwagen and Orians in S. R. Kellert and E. O. Wilson, eds., *The Biophilia Hypothesis* (Washington, D.C.: Island Press/Shearwater Books, 1993); and Gordon H. Orans, *Bulletin of the Ecological Society of America* 79 (1): 15–28 (1998).

136   I have borrowed the image of the **seventy-year life span** to scale down the history of *Homo* from Howard Frumkin, *American Journal of Preventive Medicine* 20 (3): 234–40 (2001).

137   Details of the **childhood development of biophilia and habitat preference** are reviewed by Roger S. Ulrich in S. R. Kellert and E. O. Wilson, eds., *op. cit.;* and Peter H. Kahn Jr. in *Developmental Review* 17 (1): 1–61 (1997) and *The Human Relationship with Nature: Development and Culture* (Cambridge: MIT Press, 1999).

138   **Childhood's hideaways** are described by David T. Sobel in *Children's Special Places: Exploring the Role of Forts, Dens, and Bush Houses in Middle Childhood* (Tucson: Zephyr Press, 1993), p. 90; and Will Nixon in *The Amicus Journal,* pp. 31–5 (summer 1997). My own episode is taken from the *Michigan Quarterly Review,* p. 90 (summer 2000).

140   The many **effects on health of pets and exposure to the natural envi-**

ronment are reviewed by Roger S. Ulrich et al., *Journal of Environmental Psychology* 11 (3): 201–30 (1991); R. S. Ulrich in S. R. Kellert and E. O. Wilson, eds., *The Biophilia Hypothesis* (Washington, D.C.: Island Press, 1993); Russ Parsons et al., *Journal of Environmental Psychology* 18 (2): 113–40 (1998); and Howard Frumkin, *American Journal of Preventive Medicine* 20 (3): 234–40 (2001).

141   The development of **biophobia,** especially with reference to genetically prepared learning of aversions to dangerous animals, is reviewed by Roger S. Ulrich in S. R. Kellert and E. O. Wilson, eds., *op cit.* The relation of snake aversion in particular to the evolution of culture was first developed by Balaji Mundkur in *The Cult of the Serpent: An Interdisciplinary Survey of Its Manifestations and Origins* (Albany: State Univ. of New York Press, 1983) and elaborated by E. O. Wilson in *Biophilia* (Cambridge, MA: Harvard Univ. Press, 1984).

143   There is a large, thoroughgoing, and mostly American literature on the **idea of wilderness.** Among the sources I have consulted are Roderick Nash, *Wilderness and the American Mind,* third ed. (New Haven: Yale Univ. Press, 1982); Bill McKibben, *The End of Nature* (New York: Random House, 1989); Frans Lanting and Christine K. Eckstrom, *Forgotten Edens: Exploring the World's Wild Places* (Washington, D.C.: National Geographic Society, 1993); J. Baird Callicott et al., "A Critique and Defense of the Wilderness Idea," a special section of *Wild Earth,* pp. 54–68 (winter 1994/95); David R. Brower, with Steve Chapple, *Let the Mountains Talk, Let the Rivers Run: A Call to Those Who Would Save the Earth* (San Francisco: HarperCollins, 1995); Lawrence Buell, *The Environmental Imagination: Thoreau, Nature Writing, and the Formation of American Culture* (Cambridge, MA: Belknap Press of Harvard Univ. Press, 1995); William Cronon, ed., *Uncommon Ground: Toward Reinventing Nature* (New York: W. W. Norton, 1995); Tom Petrie, Kim Leighton, and Greg Linder, eds., *Temple Wilderness: A Collection of Thoughts and Images on Our Spiritual Bond with the Earth* (Minocqua, WI: Willow Creek Press, 1996).

CHAPTER 7: THE SOLUTION

150   The **income disparities** of the richest and poorest countries are cited from the United Nations' *Human Development Report 1999* and discussed by Fouad Ajami in *Foreign Policy* 119: 30–4 (summer 2000). The consequences of the disparity are explored by Geoffrey D. Dabelko in the *Wilson Quarterly* 23 (4): 14–19 (autumn 1999) and by Thomas F. Homer-Dixon in *Environment, Scarcity, and Violence* (Princeton: Princeton Univ. Press, 1999) and *The Ingenuity Gap* (New York: Knopf, 2000).

150   On the **difference in consumption** by rich and poor nations: William E. Rees and Mathis Wackernagel in AnnMari Jansson et al., eds., *Investing in*

Natural Capital: The Ecological Economics Approach to Sustainability (Washington, D.C.: Island Press, 1994), pp. 362–90. The four-worlds estimate is from a personal communication from Mathis Wackernagel (January 24, 2000) (Redefining Progress, 1 Kearny St., San Francisco, CA); see the explanation of the concept of the ecological footprint in chapter 2 of the present book.

155 The poll of **American attitudes toward the natural world** and the values that shape them was conducted by the research firm Belden & Russonello and Research/Strategy/Management (R/S/M), commissioned by the Communications Consortium Media Center on behalf of the Consultative Group on Biodiversity, and published as a report, "Human Values and Nature's Future: American Attitudes on Biological Diversity" (October 1996). The results are cited here by permission of the CCMC.

157 Reports on the **Christian and Jewish environmental action groups,** with interviews of some of their leaders, are provided by Caryle Murphy in the *Washington Post,* pp. A1–6 (February 3, 1998); and Michael Paulson in the *Boston Globe,* p. B3 (October 14, 2000). A general account of the **relation of conservation to faith in religions generally** is provided by Libby Bassett, John T. Brinkman, and Kusimita P. Pedersen, eds., *Earth and Faith: A Book of Reflection for Action* (New York: United Nations Environment Program, 2000). **Janisse Ray warns the loggers** of God's displeasure in *Ecology of a Cracker Childhood* (Minneapolis, MN: Milkweed Editions, 1999).

159 The statement of principle of the **Religious Campaign for Forest Conservation** was given by Fred Krueger in *Religion and the Forests* 1 (1): 2 (spring 2000).

160 The twenty-five **terrestrial hotspots** listed by Norman Myers and subsequently refined by Myers with Russell Mittermeier, Gustavo Fonseca, and other staff of Conservation International (*Nature* 403: 853–8, 2000) are the tropical Andes; Mesoamerica (southern Mexico to Costa Rica); the Caribbean islands; the Brazilian Atlantic forest; Panama and Chocó of Colombia to western Ecuador; the cerrado (savanna) of Brazil; central Chile; California Floristic Province (coastal Mediterranean scrubland); Madagascar; the eastern mountain arc and coastal forests of Tanzania and Kenya; the West African forests; Cape Floristic Province of South Africa; Succulent Karoo of South Africa; the Mediterranean Basin rim; the Caucasus region; Sundaland (the great islands of Indonesia and surrounding shelf islands); Wallacea (the Lesser Sunda Islands of Indonesia, Lombok to Timor); the Philippines; Indo-Burma; south-central China; Sri Lanka and the Western Ghats of India; southwestern Australia (Mediterranean-climate scrubland); New Caledonia; New Zealand; Polynesia plus Micronesia. Each area is regarded as a hotspot either in entirety or as fragments

contained within them. A beautifully illustrated description of these critical areas is provided in *Hotspots: Earth's Biologically Richest and Most
Endangered Terrestrial Ecoregions,* by Russell A. Mittermeier, Norman
Myers et al. (Mexico City: CEMEX, Conservation International, 1999).
The staff of the World Wildlife Fund has independently defined Global
2000 Ecoregions, covering both terrestrial and marine environments,
within which hotspots can be more precisely pinpointed; their databases
and recommended conservation are spelled out in the WWF annual
reports and attendant publications (www.worldwildlife.org). The terrestrial hotspots defined by Conservation International and World Wildlife
Fund respectively have a more than 80 percent overlap.

160   Many of the **special recommendations developed by biologists and
environmental scientists for conserving biodiversity** while enhancing
agriculture, forestry, and the general economy are spelled out in my earlier
work *The Diversity of Life* (Cambridge, MA: Belknap Press of Harvard
Univ. Press, 1992; paperback, with college textbook addendum by Dan L.
Perlman and Glenn Adelson, New York: W. W. Norton, 1993); as well as
standard conservation textbooks and practical guides such as John F.
Ahearne, H. Guyford Stever, et al., *Linking Science and Technology to Society's Environmental Goals* (Washington, D.C.: National Academy Press,
1996); William J. Sutherland, ed., *Conservation Science and Action* (Mal-
den, MA: Blackwell Science, 1998); W. L. Sutherland, *The Conservation
Handbook: Research, Management and Policy* (Malden, MA: Blackwell Sci-
ence, 2000); Michael E. Soulé and John Terborgh, eds., *Continental Con-
servation: Scientific Foundations of Regional Reserve Networks* (Washington,
D.C.: Island Press, 1999); Donald Kennedy and John A. Riggs, eds., *U.S.
Policy and the Global Environment: Memos to the President* (Washington,
D.C.: Aspen Institute, 2000); and Peter H. Raven, ed., *Nature and
Human Society: The Quest for a Sustainable World* (Washington, D:C.:
National Academy Press, 2000). Estimates of the pressure of population
pressure and rising crop production on natural environments are reviewed
by David Tilman, *Proceedings of the National Academy of Sciences, USA*
96: 5995–6000 (1999).

164   For the idea of a **one-cent coffee tax,** I am grateful to Daniel H. Janzen.

165   A complex, cross-indexed, and annotated **catalog of governmental and
nongovernmental conservation organizations** and nature reserves of
North America is provided in the annual editions of the *Conservation
Directory,* published by the National Wildlife Federation (http://www.nwf.
org/nwf).

165   The growing **number of nongovernment organizations** devoted to
humanitarian and environmental problems is taken from the *Yearbook of
International Organizations 1996–1997* (Munich: K. G. Saur Verlag, 1997)

and cited in an analysis of information technologies and environmental studies by Molly O'Meara in *State of the World 2000* (New York: Norton/Worldwatch Books, 2000).

166 The **percentage of people belonging to environmental organizations** is reported by Norman Myers in *BioScience* 49 (10): 834–5, 837 (October 1999).

166 On the **assets of the largest corporations:** see Paul Hawken in *World•Watch* 13 (4): 36 (2000).

168 The contrasting views of the **Cameroonian journalist and the president of the World Wide Fund for Nature** concerning the logging of the African forests are reported in the *Economist* 351: 54–5 (June 26, 1999).

169 The size of **memberships of most of the leading conservation groups** was cited in *Marketing as a Conservation Strategy,* a booklet published by The Nature Conservancy (www.tnc.org) in 1999. The estimate for the World Wildlife Fund (WWF) was provided by Kathryn S. Fuller and James P. Leape of WWF (personal communication).

170 The $1 billion **Campaign for Conservation** by The Nature Conservancy was announced by its president, John C. Sawhill, in *Nature Conservancy*, p. 5 (May/June 2000), and described in the lead editorial of the *New York Times* (March 17, 2000). The **land acquisition program** of the campaign is based upon TNC's long-standing natural heritage programs, recently reorganized as part of the independent Association for Biodiversity Information. Some of the databases are summarized in *Precious Heritage: The Status of Biodiversity in the United States,* edited by Bruce A. Stein, Lynn S. Kutner, and Jonathan S. Adams (New York: Oxford Univ. Press, 2000).

170 The **role of the World Wildlife Fund in establishing Amazonian parks** is described by Lesley Alderman in *Barron's National Business and Financial Weekly,* pp. 22–3 (December 18, 2000).

172 **Conservation concessions** as "warp speed" conservation: Richard Rice (personal conversation).

173 The **Guyana forest concessions** purchased by Conservation International and its partners are described in the *Global Environmental Change Report* 12 (19): 1–2 (2000), as well as by Reed Abelson in the *New York Times,* Business World (September 24, 2000). I have also drawn upon press releases and internal reports provided by Conservation International (http://www.conservation.org).

174 The purchase of **logging rights in Bolivia** by The Nature Conservancy and Conservation International, used to increase the size of the Noel Kempff Mercado and Madidi National Parks, is reported by R. E. Gullison, R. E. Rice, and A. G. Blundell in *Nature* 404: 923–4 (2000).

174 My account of the **Suriname Conservation Foundation** and establish-

ment of the trust fund to support forest conservation in that country is based on press releases and internal reports of Conservation International (http://www.conservation.org), including a booklet, *The Central Suriname Nature Reserve* (2000), as well as personal accounts by Russell A. Mittermeier, Conservation International's president.

178   **The Wildlands Project** and its rationale are described by Michael E. Soulé and John Terborgh, *BioScience* 49 (10): 809–17 (1999); David Foreman, *Denver University Law Review* 76 (2): 535–55 (1999); Jocelyn Kaiser, *Science* 289: 2259 (2000); and, most comprehensively, the special issue of *Wild Earth* (10: 1, 2000) devoted to the wildlands concept and its various ramifications.

179   The record of **land acquisition by The Nature Conservancy** is based on an internal memorandum by the organization's president, John C. Sawhill (October 26, 1999).

179   The estimates of **private property ownership** provided here were made by J. Michael Scott of the University of Idaho (personal communication, June 28, 1999). He based his calculations in part on the assessment by James A. Lewis, *Landownership in the United States in 1978* (Agriculture Information Bulletin no. 435) (Washington, D.C.: U.S. Department of Agriculture, 1980). The one hundred owners with the largest private U.S. holdings in 1997 are described in *Worth*, pp. 78–89 (February 1997).

180   The purchase of the Pacific island of **Palmyra** by The Nature Conservancy for $37 million is described in the organization's journal, *Nature Conservancy*, p. 29 (January/February 2001), and in the lead editorial of the *New York Times* on June 11, 2000. The purchase of the **Cuatro Ciénagas** oasis by the same organization was announced in *Nature Conservancy*, p. 28 (May/June 1998).

181   The role of **private nature reserves in Costa Rica** is analyzed by Patrick Herzog and Christopher Vaughan in *Revista de Biología Tropical* 46 (2): 183–9 (1998).

181   The **advantages of private enterprise**, including nongovernmental organizations (NGOs), in global conservation have been discussed in a substantial literature during the past decade, which is succinctly summarized by Gretchen C. Daily and Brian H. Walker, *Nature* 403: 243–5 (2000). Strong support for the engagement of private enterprise comes from NGOs all across the political spectrum—from the liberal Natural Step (www.emis.com/tns), for example, to the conservative Political Economy Research Center (PERC) (perc@perc.org). The essential **role of government** in addition to private enterprise is stressed by Alexander James, Kevin J. Gaston, and Andrew Balmford in *Nature* 404: 120 (2000).

182   Alexander James et al. *(ibid.)* estimate that the **annual cost of global conservation** sufficient to sustain a representative sample of Earth's eco-

systems, might be "around $27.5 billion." The rough estimates of the costs of maintaining adequate reserves of tropical forests is from the 2000 conference "Defying Nature's End," reported by Stuart L. Pimm et al., *Science* 293: 2207–8 (2001).

183 **Perverse subsidies** as a major burden on both the economy and the environment are analyzed by Norman Myers and Jennifer Kent in *Perverse Subsidies: How Tax Dollars Can Undercut the Environment and the Economy* (Washington, D.C.: Island Press, 2001), and a brief overview of the subject is given by Myers in *Nature* 392: 327–8 (1998). See also *Paying the Piper: Subsidies, Politics, and the Environment* (Worldwatch Paper no. 133), by David Malin Roodman (Washington, D.C.: Worldwatch Institute, 1996). A special study of the effect of such subsidies on forests, with a critique of the plan adopted for forest protection by the Group of Eight (G8) countries (Canada, France, Germany, Italy, Japan, Russia, the United Kingdom, and the United States) is provided by Nigel Sizer et al. in the June 2000 *Forest Notes* of the World Resources Institute.

185 There is a veritable library of reports and analyses on the **Convention on Biological Diversity** of the United Nations Conference on Environment and Development (UNCED, or the Earth Summit), held in Rio de Janeiro during two weeks in June 1992. See, for example, *The Earth Summit: A Planetary Reckoning,* by Adam Rogers (Los Angeles: Global View Press, 1993).

185 The proposed **Korean demilitarized zone (DMZ) biodiversity reserve** is described by Ke Chung Kim in *Science* 278: 242–3 (1997); it is being promoted (in 2001) by the DMZ Forum, sponsored by multiple environmental organizations (http://dmz.koo.net).

185 An authoritative account of the content and history of the **Endangered Species Act** of the United States is offered by Michael J. Bean in *Environment* 41 (1): 12–18, 34–8 (1999). For parallel read-ing, with examples and personal anecdotes, see Douglas H. Chadwick, *National Geographic* 187 (3): 2–41 (March 1995). A brief but excellent chronicle of key events in the history of the Ameri-can environment movement are given in an essay on wilderness by Stewart L. Udall in *American Heritage*, pp. 98–105 (February/March 2000). Practical applications, including the use of habitat conservation plans, are summarized by Laura C. Hood et al., *Frayed Safety Nets: Conservation Planning Under the Endangered Species Act* (Washington, D.C.: Defenders of Wildlife, 1998).

GLOSSARY

*In this short collection I have included terms of a more technical
nature that may not be immediately familiar to all readers.*

adaptive radiation The splitting of a single species into many species that
occupy diverse ways of life within the same geographic range. The classical
example is the proliferation of the Australian marsupial mammals from a single
distant ancestor into kangaroos, wombats, phalangers, and other mammalian
forms with analogs elsewhere in the world.

archaean A member of the Archaea, a distinct, bacterialike kingdom of single-
celled organisms, often found in extreme habitats such as hot springs but also in
the open sea and other more "normal" environments.

area-species principle The arithmetically regular relation between the area of an
island or habitat and the number of species that live sustainably within it.

autotroph Any organisms able to live and reproduce without eating other
organisms or their parts: in particular, plants, which utilize energy from sunlight, and microorganisms that obtain energy from the oxidation of inorganic
molecules.

biodiversity Biological diversity for short. All of the hereditary variation in
organisms, from differences in ecosystems to the species composing each ecosystem, thence to the genetic variation in each of the species. As a term, biodiversity

may be used to refer to the variety of life of all of Earth or to any part of it—hence the biodiversity of Peru or the biodiversity of a Peruvian rainforest.

biophilia The innate tendency to be attracted by other life forms and to affiliate with natural living systems.

biosphere All of life, hence all of living plants, animals, and microorganisms. As a diaphanous film enveloping the planet, the biosphere forms a hollow sphere.

biota All the different kinds of organisms—plants, animals, and microorganisms—found in a particular place.

bottleneck A constricted region through which some entity, such as a liquid, a population, or diversity, must pass to reacquire, or at least approach, its original condition. The human bottleneck of the twenty-first century was created by increasing population and per-capita consumption and decreasing per-capita nonrenewable natural resources.

chromosome In higher organisms (i.e., other than bacteria and archaeans), a segment made up of genes and the protein surrounding them.

coevolution Evolution in two (or more) species of organisms such that changes in one affect changes in the other(s).

Conservation International (CI) A global conservation organization headquartered in Washington, D.C.

Darwinism Evolution by natural selection, named after the discoverer of this fundamental natural process, Charles Darwin.

DNA (deoxyribonucleic acid) The helical chain molecule that composes the genetic code.

ecological footprint The amount of productive land appropriated on average by each person for food, water, transportation, habitation, waste management, government, and entertainment.

ecology The scientific study of the interaction of organisms with their environment, including the physical environment and the other organisms in it.

ecosystem The physical environment plus the organisms living in it of a particular habitat, such as a forest or a coral reef (or, in increasing scale, up to the whole planet). Ecosystems can be natural or artificial.

ecosystem services The role played by ecosystems in creating a healthful environment for human beings, from production of oxygen to soil genesis and water detoxification.

ecotourism Tourism focused on attractive and interesting features of the environment, including the fauna and flora.

energy pyramid Each trophic level, from plants to plant eaters to carnivores, converts very roughly 10 percent of the energy it obtains into its own tissue. The resulting decline in energy available to each higher trophic level in turn creates the energy pyramid.

exponential Increasing or decreasing by compound growth or compound reduction. Both a population and a bank account grow exponentially if undisturbed. A population or bank account also declines exponentially if it is diminished at regular intervals by a fixed percent of the amount still present.

extremophiles Organisms, usually bacteria and archaeans but also a few algae, fungi, and invertebrates, that are adapted to live in extreme environmental conditions such as hot springs, slurried marine ice, and deep subterranean crevices.

fauna All the animals found in a particular place.

flora All the plants found in a particular place.

gene The fundamental hereditary unit, composed of multiple base pairs (of DNA) and usually located within a short segment of a chromosome.

genome All the genes of a particular organism or of a species.

genus (pl. genera) An assemblage of similar and closely related species.

habitat An environment of a certain kind, such as a lake or a glade in a forest.

hotspot A region of the world, such as Madagascar and the tropical Andes, that is both rich in species found nowhere else and environmentally endangered.

invasive species A species of plant, animal, or microorganism that is both alien to the environment in which it lives and destructive in some manner to the environment and its inhabitants.

invertebrates Animals lacking a backbone of bony segments that enclose the central nerve cord. Most animals are invertebrates, from roundworms to starfish, and from insects to clams.

IUCN The International Union for the Conservation of Nature and Natural Resources (also known as the World Conservation Union), headquartered in Gland, Switzerland.

mammals Animals of the class Mammalia, characterized by milk produced in the female mammary gland and by body hair.

megafauna The largest animals, weighing ten kilograms or more, such as ostriches, deer, and crocodiles.

microbev A microorganism, especially a bacterium or archaean.

microorganism An organism too small to be seen with the naked eye, typically a bacterium, archaean, or protistan (protozoan), but also any one of the smallest forms of fungi and algae.

natural selection The differential contribution of offspring to the next generation by various genetic types belonging to the same population: the mechanism of evolution suggested by Charles Darwin.

NGO See *nongovernmental organization.*

nongovernmental organization (NGO) An organization that operates separately from national and local governments.

order of magnitude To the nearest power of ten. The number 769 is, for example, closer to the third power of ten, or 1,000; the number 461 is closer to the second power of ten, or 100.

phylogeny The evolutionary history of a particular group of organisms, such as orchids or swallowtail butterflies, with special reference to the family tree of the species belonging to the group.

plankton Organisms that float passively in the sea and air, comprising mostly microorganisms together with small plants and animals.

race See *subspecies.*

rainforest A forest with rainfall sufficiently abundant and evenly enough distributed through the year to sustain dense evergreen trees. The most familiar and biologically diverse of such ecosystems are the tropical rainforests, which are typically structured into several irregular canopy layers thick enough to capture 90 percent of the sunlight before it reaches the ground. Temperate rainforests also exist, for example in the Pacific Northwest of North America, the southern coast of Chile, and Tasmania.

Red List The global lists of threatened animal and plant species assembled and published by the Species Survival Commission of the IUCN – World Conservation Union. The most recent lists were published in 2000.

species The basic unit of classification, consisting of a population or series of populations of closely related and similar organisms. In sexually reproducing organisms, the species is usually more narrowly defined by the biological species concept: to wit, a population or series of populations of organisms that under natural conditions freely interbreed with one another but not with members of other species.

subspecies A subdivision of a species. Usually defined as a geographical race: a geographically discrete population that differs in one or more hereditary traits from other geographic populations of the same species.

synergism The enhancement of the effect of two or more factors when they occur together, as for example the increase of tissue production in two or more particular plant species when they are grown together.

systematics The naming and classification of organisms. Essentially the same as taxonomy, but with an additional emphasis on the tracing of evolutionary lineages and the clustering of species into larger groups (such as genera, orders, and phyla) based on knowledge of the lineages.

taxonomy The naming and classification of organisms. Essentially the same as systematics, but usually applied to the narrower procedures of description, formal naming, and clustering of species into higher categories such as genera, orders, and phyla.

The Nature Conservancy (TNC) A conservation organization concentrated on acquisition and protection of nature reserves; mostly United States but increasingly international; headquartered in Arlington, Virginia.

transgene A gene transferred by genetic engineering from one species of organism to another.

vertebrates Animals that possess backbones of bony segments. Living vertebrates compose five major groups: fishes, amphibians (frogs, salamanders, and caecilians), reptiles, birds, and mammals.

World Conservation Union A major global organization, headquartered in Gland, Switzerland; also known as the IUCN.

WWF Acronym used both by the World Wildlife Fund, headquartered in Washington, D.C., and the World Wide Fund for Nature, located in Gland, Switzerland. The World Wildlife Fund is a national affiliate of the World Wide Fund for Nature, the latter comprising numerous such affiliates worldwide. Both WWFs are global conservation organizations of major importance.

---○---

ACKNOWLEDGMENTS

The past twenty years have witnessed the maturing of the conservation movement into a potent global enterprise. Today it is highly eclectic, having been assembled from elements of biology, economics, anthropology, political science, aesthetics, and, not least, religious belief and moral philosophy. Its driving theme is that the well-being of humanity is locked together with the health of the living planet, and that the stewardship of nature is therefore as equally important to a Manhattan banker as it is to a Honduran campesino.

I have accordingly turned to a wide diversity of consultants in composing this book. I am especially grateful to the following experts who shared important information in their fields, reviewed the relevant portions of the manuscript, or (usually) both:

Michael J. Bean (environmental law)
Andrew A. Bryant (marmot biology)
Lawrence Buell (Thoreau)
Bradley P. Dean (Thoreau)
Gustavo Fonseca (conservation biology, public policy)
Thomas J. Foose (rhinoceros biology)
Howard Frumkin (biophilia, public health)
Kathryn S. Fuller (conservation, public policy)
Ted Gullison (forest management)

James Leape (conservation, public policy)
Edward J. Maruska (rhinoceros biology)
James J. McCarthy (climate)
Russell A. Mittermeier (conservation biology, public policy)
Norman Myers (conservation biology)
Brad Parker (Thoreau)
Stuart L. Pimm (conservation biology)

Alan Rabinowitz (rhinoceros biology)

Richard Rice (ecological economics)

Terri Roth (rhinoceros biology)

Stuart L. Schreiber (natural products)

J. Michael Scott (conservation, land management)

Peter A. Seligmann (conservation, public policy)

M. S. Swaminathan (agriculture)

David Tilman (ecosystems research)

Mathis Wackernagel (global natural resources)

Diana H. Wall (biodiversity, Antarctica)

Christopher T. Walsh (natural products, biomedicine)

Of course, all are exonerated from errors and misconceptions remaining as the manuscript goes to press (August 2001).

Finally, as has been the case since my first book, *The Theory of Island Biogeography,* coauthored with Robert H. MacArthur, appeared in 1967, the present work has been steered by the unerring good advice and expert assistance of Kathleen M. Horton. It is a pleasure to acknowledge her special role, past and present, in research and editing.

INDEX

Edward O. Wilson was born in Birmingham, Alabama, in 1929. He received his B.S. and M.S. in biology from the University of Alabama and, in 1955, his Ph.D. in biology from Harvard, where he taught until 1997, and where he received both of its college-wide teaching awards. He is currently Pellegrino University Research Professor and Honorary Curator in Entomology of the Museum of Comparative Zoology at Harvard. He is the author of two Pulitzer Prize–winning books, *On Human Nature* (1978) and *The Ants* (1990, with Bert Hölldobler), as well as the recipient of many fellowships, honors, and awards, including the 1976 National Medal of Science; the Crafoord Prize from the Royal Swedish Academy of Sciences (1990); the International Prize for Biology from Japan (1993); the King Faisal International Prize for Science from Saudi Arabia (2000); and, for his conservation efforts, the Gold Medal of the World Wide Fund for Nature (1990) and the Audubon Medal of the National Audubon Society (1995). He is on the boards of directors of The Nature Conservancy, Conservation International, and the American Museum of Natural History, and gives many lectures throughout the world. He lives in Lexington, Massachusetts, with his wife, Renee.

A NOTE ON THE TYPE

This book was set in Adobe Garamond. Designed for the Adobe Corporation by Robert Slimbach, the fonts are based on types first cut by Claude Garamond (c. 1480–1561). Garamond was a pupil of Geoffrey Tory and is believed to have followed the Venetian models, although he introduced a number of important differences, and it is to him that we owe the letter we now know as "old style." He gave to his letters a certain elegance and feeling of movement that won their creator an immediate reputation and the patronage of Francis I of France.

Composed by Creative Graphics, Allentown, Pennsylvania
Printed and bound by Berryville Graphics, Berryville, Virginia
Designed by Robert C. Olsson